NEW POLYMERIC MATERIALS
Reactive Processing and Physical Properties

ORGANIZING AND SPONSORING INSTITUTIONS

European Physical Society (EPS)

European Polymer Federation (EPF)

Italian National Research Council (CNR)

Italian Society of Science and Technology of
Macromolecules (AIM)

D.S.M. Research and Patent (Holland)

Institute of Research on Polymer Technology of
Italian CNR (Naples)

Studies Center for Synthetic and Natural
Macromolecules of Italian CNR (Genoa)

> CHAIRMAN
> Ezio Martuscelli
> Institute of Research on Polymer
> Technology of Italian CNR
> Via Toiano, 6 - Arco Felice
> 80072 Naples - Italy

CO-SPONSORING INSTITUTIONS

3 M Italia

Alusuisse Italia

Ausimont

Enichem Elastomeri

Himont

Industrie Pirelli

Pomini Farrel

Savid

Sir

Tecnopolimeri

NEW POLYMERIC MATERIALS

Reactive Processing and Physical Properties

Proceedings of the International Seminar
9-13 June 1986, Naples, Italy

Edited by

E. Martuscelli and C. Marchetta

*Institute of Research on Polymer Technology
of Italian CNR (Naples)*

CRC Press
Taylor & Francis Group
Boca Raton London New York

CRC Press is an imprint of the
Taylor & Francis Group, an **informa** business

VNU Science Press BV
P.O. Box 2093
3500 GB Utrecht
The Netherlands

© 1987 VNU Science Press BV

First published in 1987

CIP-DATA Koninklijke Bibliotheek, Den Haag

New polymeric materials: reactive processing and physical
properties / ed. E. Martuscelli and C. Marchetta.
Utrecht: VNU Science Press. - III.
Invited papers presented at a symposium held 9-13 june
1986 at Naples, Italy.
ISBN 90-6764-091-3 bound
SISO 542.5 UDC 66.095.2
Subject heading: polymeric materials.

CONTENTS

Preface

This volume contains invited papers presented at the International Symposium on "New Polymeric Materials: Reactive Processing and Physical Properties" held in Naples, Italy, from June 9 to 13 1986. Approximatively 150 participants from fifteen countries were present. The scientific programme included morning and evening lectures by invited speakers and poster sessions in the late afternoon with a total of 50 posters being presented.

The field of reactive polymer processing traditionally included the thermosetting and elastomeric systems. The continuous growth and the great effort made in the formulation of new materials, highly competitive with conventional materials in terms of costs and performances, have expanded this field considerably. The most interesting developments have occurred in RIM, R-RIM, grafting and functionalization, reactive blending, INP and S-INP, SMC and BMC.

The aim of the conference was to discuss recent developments in the field and future trends. Special attention was given to the chemical, kinetic and rheological aspects of the reacting systems and to relationships between the parameters regulating the reactive processing and the physical properties of the resulting materials.

The programme was concluded by a discussion, led by a panel of leaders in this field, of unresolved problems and of emerging new areas for research. This stimulating programme was made even more exciting by the combination of an excellent audience.

E. Martuscelli
(Chairman)

C. Marchetta
(Member of Organizing Committee)

PHASE SEPARATION PROFILES IN RUBBER MODIFIED THERMOSETS OBTAINED BY REACTIVE PROCESSING

R.J.J.Williams , H.E.Adabbo , A.J.Rojas and J.Borrajo

Institute of Materials Science and Technology (INTEMA), University of Mar del Plata and National Research Council (CONICET), J.B.Justo 4302 , (7600)Mar del Plata, Argentina

ABSTRACT

The cure of a rubber modified epoxy resin with a diamine, in a heated mold, is analyzed through the numerical solution of the thermal energy and mass balances. The resulting conversion vs temperature trajectories, at different positions in the part, are introduced in a phase separation model previously developed, to obtain the corresponding phase separation profiles. It is shown that the concentration of dispersed phase particles as well as the volume fraction of the dispersed phase, go through a maximum located near the surface. In this region the particle size distribution is enriched in small particles. The effect of varying the wall temperature or the initial rubber concentration, upon the phase separation profiles, is discussed.

INTRODUCTION

Rubber modified thermosets are prepared from an homogeneous solution of an elastomer in a thermosetting resin, which in the course of polymerization precipitates a discrete, randomly dispersed rubbery phase. The presence of the second phase introduces a toughening effect, increasing the energy required to maintain a given crack - growth rate. Figure 1 illustrates the phase separation taking place during the cure of a rubber modified epoxy with a diamine. Initially the system is homogeneous due to the significant contribution of the entropy of mixing of small molecules to the free energy of the system. However, as crosslinking takes place the molar volume of the thermoset increases leading to a decrease in the absolute value of the entropy

1

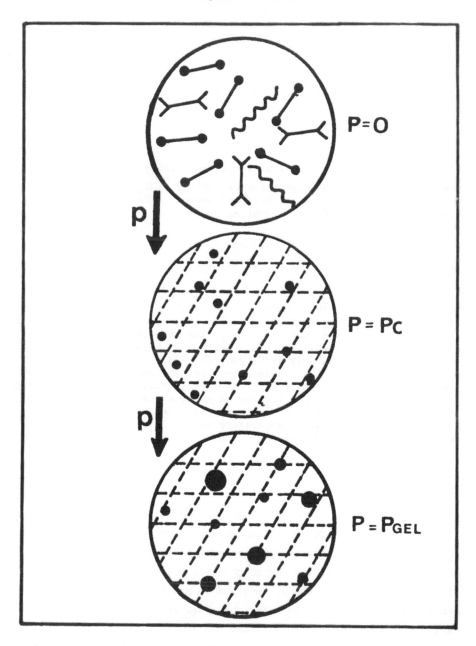

Figure 1. Phase separation during the cure of a rubber modified epoxy with a diamine (●—● : epoxy resin, ＞—＜ : diamine, ⋀⋀ : rubber). p = thermoset conversion; p_c = cloud point; p_{gel} = gel point.

of mixing. Eventually a point is reached $(p_c =$ cloud point) where
phase separation begins to take place. When the matrix gels phase
separation is prevented due to diffusional restrictions and the final
particle size distribution is attained.

Fracture mechanics studies give us certain hints regarding the
desired morphology of these materials. For example, it is known that
the interfacial adhesion between the rubber and the matrix must be
promoted through a chemical bond (i.e. end-capped rubber with the
epoxy resin); also, a high volume fraction of dispersed phase is
desired provided that the thermoset remains as the continuous phase
and the rubber is effectively depleted from the matrix. Moreover,
the particle size distribution must include enough particles of a
critical size (i.e. in the 0.5 - 5μm range) leading to localized
cavitation, microvoid development and similar mechanisms. Also, the
presence of very small particles (i.e. in the 0.01÷0.03 μm range)
may be useful to promote shear banding. On the other hand, when a
part made from the rubber modified thermoset is cured in a heated
mold (using RIM or a similar technique), phase separation profiles
will be developed due to changes in the conversion vs temperature
history with position in the specimen. Although it is difficult to
say anything about an optimum phase separation profile in the part,
one may infer that at and near the surface it is desirable to have
a well developed phase separation (i.e. large volume fraction of
dispersed phase, exhaustion of rubber from the thermosetting matrix,
enough concentration of particles of critical sizes). This will
produce a toughening of the surface region increasing the capability
of withstanding possible impacts.

The aim of this study is to discuss the influence of operation
variables upon the resulting morphology of parts cured in heated
molds. The cure of a rubber modified epoxy resin in a heated mold
will be analyzed through the numerical solution of thermal energy
and mass balances. The resulting conversion vs temperature trajectories, at different positions in the part, will be then introduced
into a phase separation model previously developed (R.J.J.Williams
et al.,1984; A.Vázquez et al.,to be published). The resulting phase
separation profiles will be related to the initial rubber amount

and the wall temperature selected to cure the part.

CURING OF PARTS IN HEATED MOLDS

The thermosetting system is assumed to be a low molecular weight bis-
phenol A-type epoxy resin (BADGE) modified with a carboxyl-terminated
polybutadiene - acrylonitrile (CTBN), and cured with a stoichiometric
amount of ethylenediamine (EDA). Due to the high heat generation in
the neat resin it is necessary to consider the use of a 50% by weight
of an inert filler to avoid degradation reactions.

The curing kinetics of stoichiometric mixtures of BADGE with EDA
has been reported elsewhere (Riccardi et al.,1984). In the temperature range scanned in this simulation the reaction rate may be
written as

$$r^+ = dp/dt = A \ (1 - p)^2 \ exp(-E/RT) \qquad (1)$$

where p is the reaction extent (conversion), $A = 4.5 \times 10^{12} \ s^{-1}$ and
$E/R = 12330$ K.

The curing of thermosets in heated molds has been analyzed by
different authors in recent years, and a review is available (Williams
et al.,1985 and references therein). As most molded parts are thin in
one dimension, i.e. the part thickness L, only one coordinate is relevant to the analysis. As shown in Figure 2 it will be assumed that
initial temperature T_0 , and reaction extent p_0 , are uniform through-
out the part. Also, the wall temperature T_w , is assumed to remain
constant during the cure.

With the usual assumptions of constant properties of the thermosetting polymer, the thermal energy balance may be written as

$$\partial T/\partial t = \alpha \ \partial^2 T/\partial s^2 + r^+ \Delta T_{ad} \qquad (2)$$

where α is the thermal diffusivity and ΔT_{ad} is the adiabatic temperature rise given by

$$\Delta T_{ad} = (-\Delta H)/\rho \ c_p \qquad (3)$$

$(-\Delta H)$ is the reaction heat per unit volume, ρ is the mass density,
and c_p is the specific heat. Both α and ΔT_{ad} are thermal properties
of the thermosetting compound (polymer plus rubber, fillers, etc.).

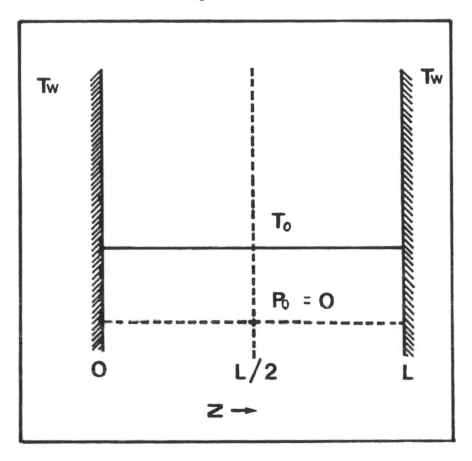

Figure 2. Schematic diagram of the mold (T_w = wall temperature, T_0 = initial temperature, p_0 = initial reaction extent, L = part thickness)

On the other hand, the balance of reactive groups may be written as

$$\partial p / \partial t = r^+ \tag{4}$$

Initial and boundary conditions are

$$T(0,z) = T_0 \ , \ p(0,z) = 0$$
$$T(t>0,0) = T_w \ , \ T(t>0,L) = T_w \tag{5}$$

By defining

$$z^* = z/L \ , \ t^* = A \ exp(-E/RT_0) \ t \ , \ T^* = (T - T_0)/\Delta T_{ad} \ ,$$

equations (2),(4) and (5) may be written as

$$\partial p / \partial t^* = r^+ A^{-1} \exp(E/RT_0) = (1 - p)^2 \exp[CT^*/(D + T^*)] \quad (6)$$

$$\partial T^* / \partial t^* = B \, \partial^2 T^* / \partial x^{*2} + (1 - p)^2 \exp[CT^*/(D + T^*)] \quad (7)$$

$$T^*(0,x^*) = p(0,x^*) = 0$$

$$T^*(t^* > 0,0) = T^*(t^* > 0,1) = F \quad (8)$$

The dimensionless parameters characterizing the curing process are:

B : $(\alpha/L^2)/[A \exp(-E/RT_0)]$ = (heat diffusion rate)/(heat production rate);

C : E/RT_0 ;

D : $T_0/\Delta T_{ad}$;

F : $(T_w - T_0)/\Delta T_{ad}$

The following values of the dimensionless parameters will be taken: B = 100 , C = 40 , D = 2.2 and F = 0.5 or 0.6 . This corresponds roughly to the cure of a 1 cm thickness part, starting at T_0 = 308 K and using wall temperatures of 378 K (F = 0.5) or 392 K (F = 0.6). The adiabatic temperature rise of the filled thermoset is ΔT_{ad} = 140 K.

The differential equations were solved using an implicit method of finite differences. Figures 3 and 4 show, respectively, temperature and conversion profiles in a part made from a 10% CTBN - modified epoxy resin. Dashed lines indicate the beginning (reach of the binodal curve) and end (gelation) of phase separation. The beginning of phase separation (binodal curve) is established by describing the free ener-gy of the system by a Flory - Huggins equation written as a function of conversion and temperature (Williams et al.,1984; Vázquez et al., to be published). As the reactivity of primary and secondary amine hydrogens in EDA is the same (Riccardi et al.,1984), the reaction may be regarded as the ideal stepwise polymerization of an A_4 (diamine) with a B_2 (epoxy). For this case, the gelation conversion is p_{gel} = 0.577.

The process begins at the surface and extends progressively through-out the part. However, when the core has ended the phase segregation, a region close to the surface is still undergoing it. It is interest-ing to point out that profiles depicted by curve 2 are obtained for

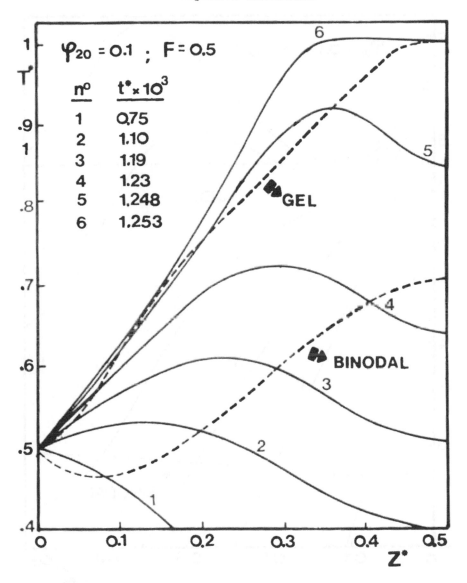

Figure 3. Temperature profiles during the cure of the rubber modified thermoset in the heated mold. Dashed lines show temperatures at which binodal curve and gelation are attained.

an actual cure time of 57.5 s, curve 4 represents the situation obtained 6.8 s later while the state of cure given by curve 6 arises

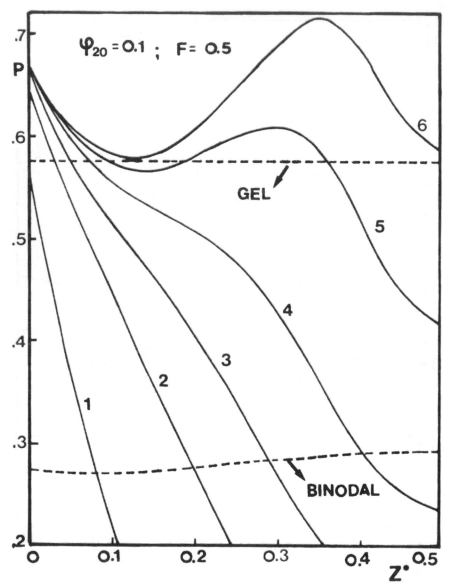

Figure 4. Conversion profiles during the cure of the rubber modified thermoset in the heated mold. Dashed lines show conversions at which binodal curve and gelation are attained.

1.2 s later than curve 4. Thus, a very rapid evolution through the metastable region is predicted near the core of the part.

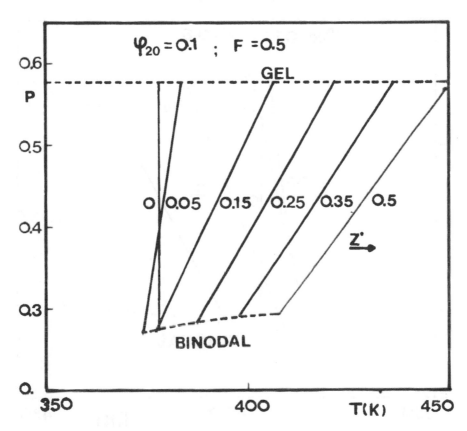

Figure 5. Conversion vs temperature trajectories for different positions in the part.

As shown by Figures 5 and 6, for two different wall temperatures (F values), conversion vs temperature trajectories in the metastable region may be approximated by straight lines. Phase separation begins at temperatures lower than the wall one in a region next to the surface and the whole trajectory may take place at $T < T_w$ (Figure 6). On the other hand, phase separation at the core proceeds at high temperatures, an effect which is enhanced by increasing the wall temperature.

The way in which these trajectories affect phase separation in the part will be analyzed in the following section.

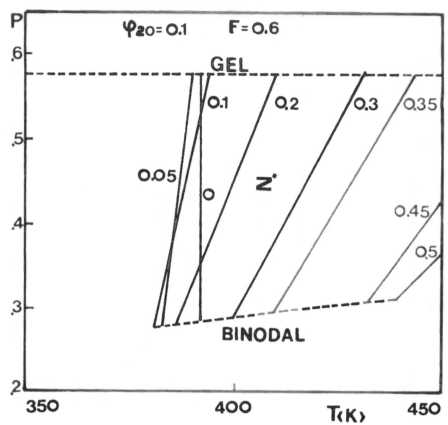

Figure 6. Conversion vs temperature trajectories for different positions in the part.

PHASE SEPARATION MODEL

A model describing thermodynamic and kinetic features of phase separation is described elsewhere (R.J.J.Williams et al.,1984; A.Vásquez et al., to be published). Figure 7 shows a block diagram of the information necessary to solve the nucleation and growth constitutive equations, and obtain parameters related to the developed morphology. P represents the concentration of particles of dispersed phase, which is used as an adjustable parameter in the simulation; $dP(r)/dr$ is the distribution function of particle sizes, V_D is the volume fraction of

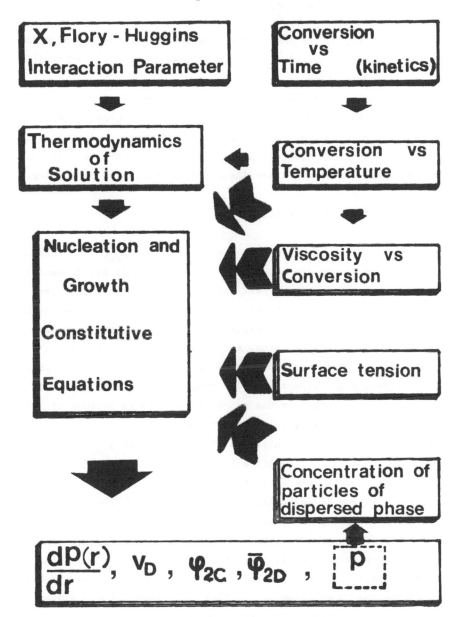

Figure 7. Block diagram of the phase separation model.

dispersed phase, ψ_{2c} is the volume fraction of rubber remaining in
the continuous phase, and $\bar{\psi}_{2D}$ is the average rubber volume fraction
of the dispersed phase.

Then, by introducing the conversion vs temperature trajectory at
a given position in the mold, the phase separation model gives us
the corresponding morphology provided that the rest of the necessary
information is available (A. Vázquez et al., to be published). The
results which are obtained are discussed in the following section.

MORPHOLOGY OF THE CURED PART

Figure 8 shows the particle size distribution, at p_{gel} , for different positions in the specimen. Most of the particles are present at
and near the surface of the part due to the low temperature levels
at which phase separation took place (less thermodynamic compatibility between rubber and thermoset), as well as the time interval
available for phase separation. Near the core, the high temperature
levels and the short residence times in the metastable region prevent
a significant phase separation. The small population of particles
with a diameter above 6 μm has not been represented because of their
limited effect upon toughening mechanisms.

Figure 9 shows the concentration of particles, at p_{gel} , as a function of position for three different curing conditions. The maximum
located near the surface is obtained for the trajectory through the
metastable region covering the lowest temperatures. The overall concentration of particles in the part is strongly dependent on the
wall temperature and to a less extent on the initial rubber concentration. Decreasing the wall temperature or increasing the rubber
amount produce less compatibility and, consequently, more separation.
An interesting observation is the fact that cases 2 and 3 showed
spinodal decomposition at the core ($z^* > 0.35$).

Figure 10 shows the volume fraction of dispersed phase, at p_{gel},
as a function of position in the part. Noteworthy, the amount of segregated phase is a strong function of the initial rubber concentration but does not show a significant dependence on the selected wall
temperature.

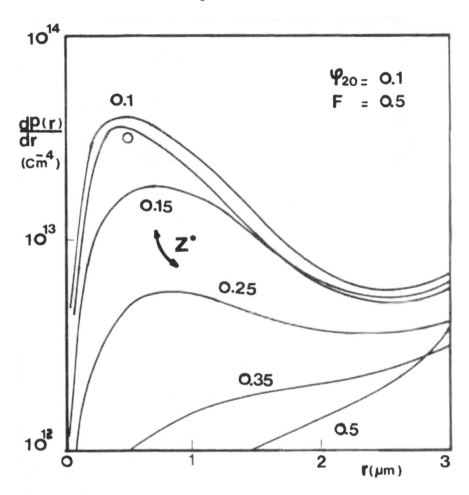

Figure 8. Particle size distribution, at p_{gel} , for different positions in the part.

By defining a volumetric average radius as

$$\bar{r} = (3 \, V_D \, / \, 4\pi P)^{1/3} \qquad , \qquad (9)$$

and using the results shown in Figures 9 and 10, one may conclude that
an increase in the wall temperature (curve 1 to curve 3) leads to a
corresponding increase in $\bar{r}$; i.e. the population is composed of less
but greater particles such that the volume fraction of dispersed phase

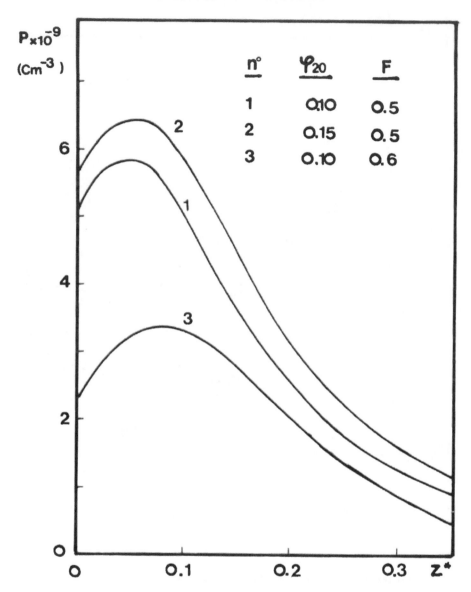

Figure 9. Concentration of dispersed phase particles, at p_{gel} , as a function of position, for different curing conditions.

is almost unchanged (at least in regions close to the surface). Also, by increasing the initial rubber concentration (curve 1 to curve 2)

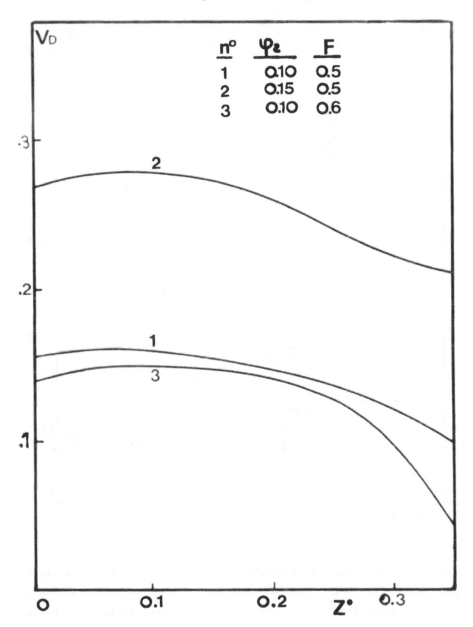

Figure 10. Volume fraction of dispersed phase, at p_gel , for different
curing conditions, as a function of position in the part.

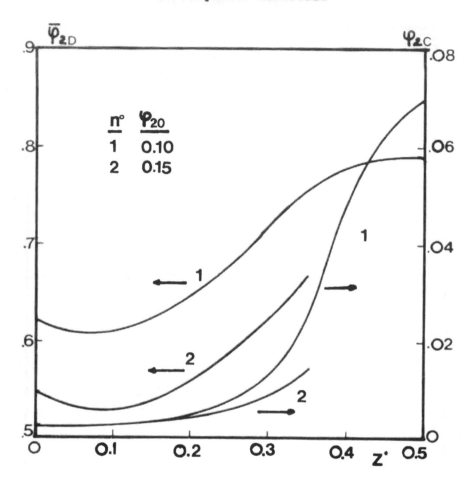

Figure 11. Final rubber concentration in the dispersed ($\bar{\varphi}_{2D}$) and continuous (φ_{2c}) phases as a function of position, for different initial rubber amounts.

there is a corresponding increase in the volumetric average radius of the population. In this case there is an increase in both number and size of dispersed particles. The prevailing effect produced by an increase in the wall temperature is the enhancement of the growth rate of dispersed particles, while the one observed when the rubber amount is increased is related to a decrease in the thermodynamic compatibility, expanding the conversion range available for phase separation.

An interesting fact arising from Figure 10 is that the volume fraction of dispersed phase, V_D , is significantly higher than the initial rubber volume fraction, φ_{20} . This means that segregated domains contain significant concentrations of the thermosetting resin. This is shown in Figure 11 for two initial rubber amounts. The segregated phase is richer in rubber when it is generated under conditions of more initial compatibility (i.e., curve 1). Except near the core of the specimen, the depletion of rubber from the thermoset is very effective for both initial rubber concentrations (i.e., very low φ_{2c} values). This means that the glass transition temperature of the material located near the surface will be almost the same as the one of the neat thermosetting resin.

Then, the results which have been discussed allow one to predict modifications of phase separation profiles in the part, arising from changes in the operation variables.

REFERENCES

Riccardi,C.C., Adabbo,H.E. and Williams,R.J.J. (1984). Curing reaction of epoxy resins with diamines. J.Appl.Polym.Sci.29, 2481 - 2492.

Vázquez,A., Rojas,A.J., Adabbo,H.E., Borrajo,J. and Williams,R.J.J. Rubber modified thermosets: prediction of the particle size distribution of dispersed domains. Polymer, to be published.

Williams,R.J.J., Borrajo,J., Adabbo,H.E. and Rojas,A.J. (1984). A model for phase separation during a thermoset polymerization. In: Rubber Modified Thermoset Resins, Riew,C.K. and Gillham,J.K.(Ed.). Advances in Chemistry Series 208, American Chemical Society, Washington D.C., pp. 195 - 213.

Williams,R.J.J., Rojas,A.J., Marciano,J.H., Ruzzo,M.M. and Hack,H.G. (1985). General trends in the curing of thermosets in heated molds. Polym.Plast.Tech.Eng.24, 243 - 266.

THERMOPLASTIC ELASTOMERS WITH GLASSY AND CRYSTALLINE DOMAINS

G. Perego*, A. Roggero*, L. Gargani**
* Eniricerche, 20097 San Donato Milanese, Milano, Italy.
** EniChem Elastomeri, 20097 San Donato Milanese, Milano, Italy.

ABSTRACT

α-methylstyrene-butadiene-α-methylstyrene (MS-B-MS) and α-methylstyrene-butadiene-pivalolactone (MS-B-PVL) linear triblocks were synthesized. Morphology of MS-B-PVL was found to be based on cylindrical microdomains containing glassy MS segments and crystalline microdomains of PVL, regularly arranged in a two-dimensional square lattice. Cylindrical microdomains in an hexagonal lattice were observed in MS-B-MS. According to dynamic mechanical and stress relaxation experiments, 'high temperature' behaviour of both copolymers is improved with respect to SBS, the best result being displayed by MS-B-PVL. Some depression of ultimate tensile properties occur in copolymers containing PVL segments.

INTRODUCTION

In a block copolymer based thermoplastic elastomer, composed of plastic and elastomeric segments, the retention of useful properties do occur at temperatures below the softening point of the plastic segments. For this reason, the use of commercially available styrene-butadiene-styrene (SBS) triblocks is practically restricted to temperatures lower than 60°C.

During the last years, a research program was undertaken in our Laboratories, with the aim of finding new materials having 'high temperature' performances improved with respect to those of SBS. In this connection, we tried to replace polystyrene by glassy segments having higher glass transition temperature (T_g).

Poly α-methylstyrene (MS) was used as the replacing glassy segment, due to its high T_g (180°C). MS-B-MS triblocks were extensively investigated, by attempting a process suitable for large volume applications (Gandini et al., 1984).

Parallely, another route was followed, by synthesizing ABC type block copolymers, in which C was a crystallizing segment. Few examples were reported in the literature concerning this type of copolymers, with the C block composed of polyethylene sulfide (Cooper et al., 1974) and polyethylene oxide (Koestier et al., 1978).

Polypivalolactone (PVL) was selected as the crystallizing segment, due to its high melting point (T_m=240°C) and to its high tendency to crystallize. Both S-B-PVL and MS-B-PVL triblocks were prepared but the attention was particularly focused on the latters, which bring together the high T_g of MS and the high T_m of PVL blocks.

The paper is dealing mainly with the results of the investigation on morphology and properties of MS-B-MS and MS-B-PVL.

RESULTS

Synthesis

The materials were synthesized by a multistep procedure. In the first step, the living MS-B diblock was prepared at room temperature. Varying amounts of unbonded poly-MS segments are formed together with the diblock.

In a second step, the MS-B-MS triblock was obtained by a coupling reaction. More details about the synthesis of these copolymers were reported elsewhere (Gandini et al., 1984).

For synthesizing MS-B-PVL triblocks, a carboxylation reaction was performed on MS-B living diblock to give the MS-B-f-COOH intermediate, according to the procedure reported by Quirck and Chen (1982). A secondary coupling reaction, leading to the formation of MS-B-MS species, occurred partially in this step.

However, the final content of MS-B-f-COOH was found to be higher than 60%.

Finally, the living polymerization of pivalolactone was carried out at 60°C on the functionalized diblock, with essentially 100% blocking efficiency and nearly 100% monomer conversion. The complete solubility of the synthesized copolymers in toluene confirms the lack of appreciable amounts of unbonded poly-PVL, the latter being insoluble in this solvent.

The molecular characteristics of the samples, determined by ^{1}H-NMR and GPC measurements, are summarized in Table 1.

Table 1
Molecular characteristics of the samples

	MS-B-PVL	MS-B-MS
X_A	0.17-0.24	0.13-0.25
$X_{A'}$	0.03-0.11	0.07-0.16
X_C	0.03-0.07	
M_A	$7.5-20 \times 10^3$	$9-14 \times 10^3$
$M_{A'}$	M_A	M_A
M_C	$2.5-8 \times 10^3$	

X_A : weight fraction of MS blocks.
$X_{A'}$: weight fraction of unbonded MS
X_C : weight fraction of PVL.
M_A : M_n of MS blocks
$M_{A'}$: M_n of unbonded MS.
M_C : M_n of PVL blocks

Morphology

Morphology was investigated mainly by small angle X-ray scattering (SAXS) and transmission electron microscopy (TEM) on compression molded samples. Ultramicrotomed thin sections, osmium tetroxide stained, were used for TEM analysis. The results concerning MS-B-MS, already reported in a paper by Gandini et al. (1984), can be summarized as in the following.

Cylindrical morphology was detected in copolymers having molecular weight   of MS block (M_A) greater than $8x10^3$, in line with what expected on the basis of their chemical composition. The structure is built up of microdomains, cylindrical in shape, formed by MS segments and dispersed in a polybutadiene matrix. Copolymers having M_A lower than $8x10^3$ display a disordered structure, constituted by irregularly shaped MS microdomains.

MS homopolymer was demonstrated to be trapped within the microdomains formed by MS blocks. We derived the simple equation (Perego et al., 1986)

$$R = R_o(1+p)q$$

relating the radius of the cylinders containing homopolymer, R, to that formed only by blocks, R_o . The parameter p represents the ratio between the content of homopolymer and that of block segments, q represents the ratio between the surface areas (average surface occupied at the interface by each covalent link between the blocks) of the domains without and with homopolymer respectively. Application of this equation with the assumption of invariant surface area (q = 1), lead to results in fair agreement with predictions of the well known Helfand's theory.

Crystalline poly-PVL in the typical α-form is clearly detected in the wide angle X-ray scattering (WAXS) pattern of MS-B-PVL. Accordingly, a melting endotherm is observed by DSC analysis, in the range 197-208°C for molecular weight of PVL block varying from 2.5 to $8x10^3$. The melting temperature remains significantly lower than that observed for high molecular weight poly-PVL (240°C). For WAXS and DSC spectra, see Figure 1.

Figure 2 shows a TEM micrograph of a typical MS-B-PVL. Elongated figures are evident, which reasonably indicate the presence of cylindrical microdomains. Moreover, a square lattice appears; the repeating unit of this lattice is formed by round

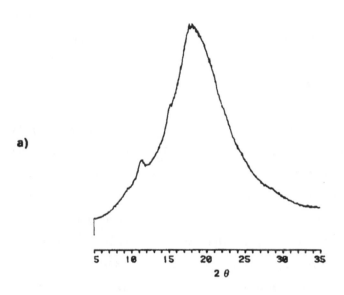

a)

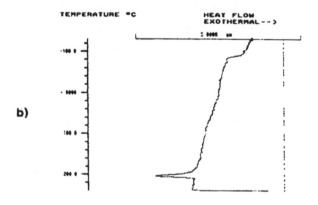

b)

Figure 1. Trace of WAXS(a) and DSC(b) patterns of MS–B–PVL
triblocks.

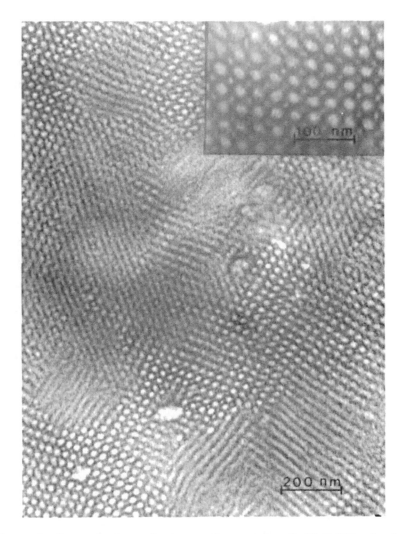

Figure 2. Transmission electron micrograph of MS-B-PVL showing
the two-dimensional square lattice morphology. The
inset shows a magnified view of the structure perpendicular to the axis of cylinders.

figures at the corners and smaller figures at the center of the square. The formers evidently constitute the cylinders viewed along their axes, the latters should represent the projection of columns of crystalline microdomains of PVL, probably oriented with the chain axis perpendicular to the axes of the cylinders. As occurs in MS-B-MS, the cylinders should contain both block and homopolymeric segments, no macroscopic segregation of homopolymer being evident from TEM analysis. Obviously, polybutadiene constitutes the continuous matrix.

Three diffraction lines are typically observed in the SAXS pattern of MS-B-PVL, with reciprocal spacings in the ratio 1:$\sqrt{2}$:2 (Figure 3a), which agree with the occurrence of a two-dimensional square lattice.

In the case of MS-B-MS, the SAXS pattern shows three lines with reciprocal spacings in the ratio 1:$\sqrt{3}$:$\sqrt{7}$, typical of a two-dimensional hexagonal structure (Figure 3b).

The occurrence of long-range structural order evidently reflects the narrow molecular weight distribution of both MS-B-MS and MS-B-PVL. It is worth to mention that the long range ordered square lattice is observed also for copolymers having M_A lower than 8×10^3. Evidently, the presence of crystalline microdomains of PVL forces the MS microdomains to assume a regular cylindrical shape and not the irregular shape occurring in the corresponding MS-B-MS triblocks.

The formation of a square lattice, instead of the hexagonal lattice, is reasonably understood by considering that the distribution of B-PVL junctions (located on two opposite sides of a crystal) doesn't fit the three-fold symmetry of the hexagonal structure.

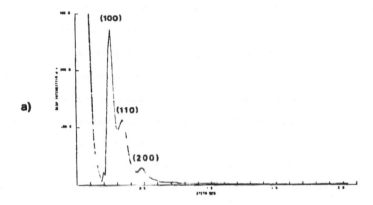

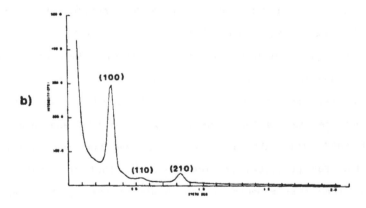

Figure 3. Trace of SAXS pattern of MS-B-PVL(a) and MS-B-MS(b)
          triblocks.

Properties

Figure 4 compares the dynamic mechanical behaviour of MS–B–MS
and SBS having similar molecular weight. Inspection of loss factor (tan δ) shows for SBS the expected transition at –80°C, characteristic of polybutadiene phase and that at ca. 100°C, characteristic of polystyrene.

The replacement of S by MS segments results in a delayed col-
lapse of the structure up to ca. 160°C, the transition of the
soft phase being practically unaffected (Figure 4a). Unfortunately, this substitution implies also an increase of dissipative behaviour in a wide temperature region, with a resulting
worse elastic behaviour at relatively low temperature. This
suggests the occurrence of a worse phase separation in the case
of MS–B–MS, which appears reasonable when considering the lower
difference between the solubility parameters of hard and soft
phases expected in this case with respect to SBS.

The elastic component of the complex modulus confirms that the
pseudo rubbery plateau of SBS is better defined, though extended
in a shorter temperature range (Figure 4b). Accordingly, stress
relaxation experiments show that SBS retains a better elastic
behaviour at room temperature while a definite improvement occurs for MS–B–MS above 60°C (Figure 5). A gain of about 20°C in
terms of service temperature can be inferred for MS–B–MS by
comparing the results for these copolymers at 80°C and those for
SBS at 60°C, the latters representing the limiting behaviour
acceptable for practical applications. The fact that this gain
is lower than that expected on the basis of the Tg of MS can be
accounted for by the apparent worse phase separation occurring
in MS–B–MS. However, the concentration and the molecular weight
distribution of homopolymer may also play some role in determin-

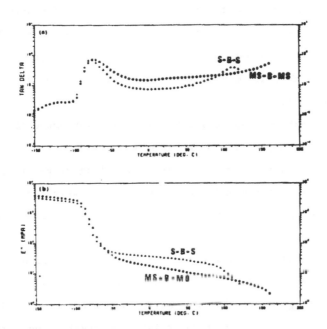

Figure 4. Loss(a) and storage(b) moduli of SBS and MS-B-MS.

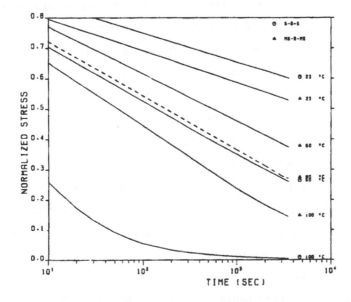

Figure 5. Stress relaxation of SBS and MS-B-MS.

ing the properties of these copolymers. The latter feature is under investigation and the results will be reported elsewhere.

Better results are apparently achieved in the presence of the crystallizing PVL segment. This is evident from Figure 6 which compares the dynamic mechanical spectra of MS-B-PVL with those of SBS. For the formers, the pseudo rubbery plateau is well defined and extended up to 170°C. The latter feature suggests an improved phase separation with respect to MS-B-MS, which seems to be in line with the tendency towards a long range structural order observed for MS-B-PVL with low molecular weight MS blocks. The definite improvement is also evident from the stress relaxation behaviour (Figure 7).

The only drawback concerns the ultimate tensile properties of MS-B-PVL, compared to those of both SBS and MS-B-MS block copolymers (Figure 8). This may be accounted for by considering the lack of anelasticity characteristic of crystalline domains, which results in a poor efficiency in dissipating energy at high deformations and, consequently, in a shorter elongation at break. Another important feature to be taken into account concerns the topology of B-PVL junctions, located on the surface of the crystals, with a spatial distribution quite different from that of MS-B junctions, located on the surface of the cylinders. Moreover, the surface area available for B-PVL junctions is certainly low, being strictly related to the unit cell dimensions of the crystals of PVL. Because of the relatively low number of chain folding expected for the crystalline PVL, due to the low molecular weight of the PVL segment, the average surface area should be well below 100 $\overset{\circ}{A}^2$ for B-PVL junctions compared to ca. 500 $\overset{\circ}{A}^2$ available for MS-B junctions. When considering these features, it is reasonable to assume the occurrence of some

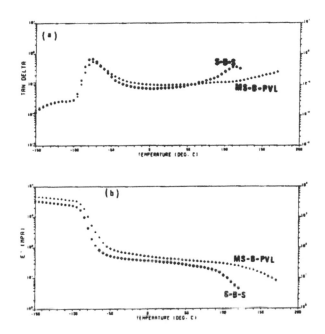

Figure 6. Loss(a) and storage(b) moduli of SBS and MS-B-PVL.

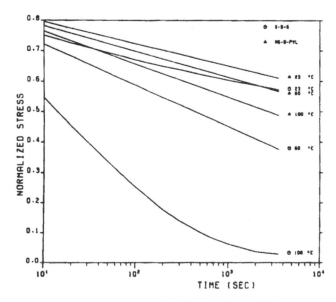

Figure 7. Stress relaxation of SBS and MS-B-PVL.

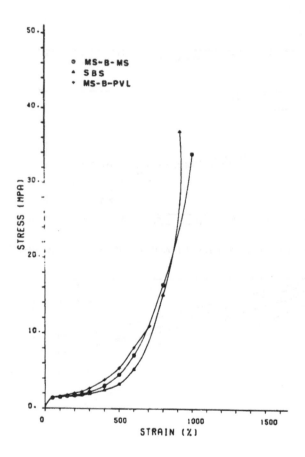

Fig. 8. Stress-strain curves of SBS, MS-B-MS and MS-B-PVL.

frozen-in deformation of rubbery segments, which could constitute a further contribution to the depressed tensile properties
of MS-B-PVL triblock copolymers.

CONCLUSIONS

α-methylstyrene–butadiene–α-methylstyrene (MS-B-MS) and α-methylstyrene–butadiene–pivalolactone (MS-B-PVL) linear triblocks have been synthesized and their morphology and properties have been investigated.

Morphology of MS-B-MS (cylindrical microdomains in an hexagonal lattice) is similar to that of SBS. MS homopolymer, which is formed during the synthesis, is dissolved within the microdomains constituted by MS blocks.

Morphology of MS-B-PVL is built up of cylindrical microdomains of MS and crystallites of PVL, regularly arranged in a two-dimensional square lattice.

Improvement of 'high temperature' performances is observed for both copolymers with respect to linear SBS. The best result is displayed by MS-B-PVL. Some depression of ultimate tensile properties occurs in copolymers containing PVL segments.

ACKNOWLEDGMENTS
The authors wish to thank Mr. E. Agostinis for his contribution in dynamic mechanical analysis, Mr. A. Gandini for his contribution in the synthesis of the materials and Mr. E. Montani for electron microscopy analysis.

REFERENCES
Cooper W., Hale T. and Walker J.S. (1974). 'Elastomeric block copolymers from ethylene sulfide'. Polymer 15, 175–186.

Gandini A., Perego G., Roggero A., Vitali R. and Zazzetta A. (1984). 'Synthesis and characterization of α-methylstyrene–butadiene–α-methylstyrene linear block copolymers'. Polymer Bulletin 12, 71–77.

Koestier D.W., Bantjes A., Feijen J. and Lyman D.J. (1978). 'Block copolymers of styrene, isoprene and ethylene oxide prepared by anionic polymerization. I. Synthesis and characterization.' J. Polymer Sci. Polymer Chem. Ed. 16, 511–521.

Perego G., Roggero A., Vitali R., Zazzetta A. and Skoulios A. (1986). 'Morphology of triblock copolymers in the presence of unbonded segments'. Makromol. Chem., submitted.

Quirck P.R. and Chen W.C. (1982). 'Functionalization of polymeric organolithium compounds. Carbonation.' Makromol. Chem.
183, 2071-2076.

TRANSESTERIFICATION REACTIONS IN MOLTEN POLYMERS

M. Lambla, J. Druz, A. Bouilloux *
ECOLE D'APPLICATION DES HAUTS POLYMERES
Institut Charles Sadron (CRM-EAHP)
4, rue Boussingault, F-67000 STRASBOURG
* ATOCHEM, CERD, F-27470 SERQUIGNY

1. INTRODUCTION :

The conversion of vinyl acetate polymers or copolymers to the corresponding polyvinylalcohol is an industrial process, applied mostly to (co)polymers containing a large amount of vinylacetate, at least 40%. These reactions, based on hydrolytic processes, are carried out in solution, emulsion or suspension, various techniques also used for the polymerization processes. This signifies that the chemical modification step can follow directly the polymerization process in the same reactor.

On the contrary, copolymers of ethylene and vinyl acetate (EVA), based on smaller amounts of the ester compound, generally from 5 to 40 % by weight, are prepared by the classical high pressure polymerization process and are recovered by extruding and pelletizing. For various reasons, essentially related to the numerous operations of dissolution or dispersion, reaction and recovery of the pure modified copolymer, it seems difficult to apply the preceeding techniques for an economic conversion of EVA copolymers. The best way to convert the ester groups in these kind of EVA copolymers is continuous modification by reactive extrusion.

Various patents report on this process based on direct hydrolysis, in presence of basic catalysts (1,2,3), but in a previous paper we mentioned the limits of these reactions and

33

introduced another chemical procedure : the transesterification of
the pendant ester groups by paraffinic alcohols introduced in the
molten polymer.

In this paper, we have studied the reactivity of various alcohols
in presence of two types of catalysts, respectively sodium
methanolate and dibutyltin dilaurate (DBTDL). The reactions are
carried on in both discontinuous and continuous mixers. The
ultimate goal of this study was to analyse the kinetic data based
on a homogeneous reaction showing that the high level of conversion
is directly connected with an equilibrated reaction of
transesterification.

2. EXPERIMENTAL :

2.1 Reagents :

Chemical reactions carried on in molten polymers may be complicated
by problems related to the limited solubility of reagents or
catalysts and, as it is generally the case in polymer modification,
by side reactions lowering the expected conversion level. For these
reasons, we have chosen the transesterification reaction for the
conversion of EVA copolymers to the corresponding EVAL (ethylene
vinylacohol copolymers). Further more we tried to use alcohol
reagents which are preferably soluble in the molten polymers. The
chosen catalysts are introduced in small amounts and are at least
soluble in the added alcohol.

On table 1, we present the main different materials : copolymers,
alcohols and catalysts which were used for the chemical reaction.

2.2. Reactive mixing :

The reaction is carried out in a single processing step, by
discontinuous, in a plastometer (Haake - Rheocord), or continuous
mixing, in a twin screw extruder (Werner - Pfleiderer).

Table 1 :   **MATERIALS**

MATERIALS	DESCRIPTION AND COMPOSITION	SOURCE	CODE
EVA Copolymers Random ethylene- vinylacetate copolymers	5 % 14 % 28 %	ATOCHEM " "	1010 VG2 1040 VNA OREVAC 9006
Methanol 1-Butanol 1-Octanol 1-Dodecanol 1-Hexadecanol 1,4-Butanediol 1,6-Hexanediol 1,12-Dodecanediol 1,2,6-Hexanetriol Trimethylolpropane Sorbitol	Commercial products (purity : 99%)	Merck Roquette	 Sorbitol
Catalyst 1 Catalyst 2	Sodium methoxyde Dibutyltindilaurate	Prolabo Merck	CH_3 ONa DBTDL

For the discontinuous process, the dried EVA resin and the chosen amounts of alcohol and catalyst are mixed before use and poured rapidly in the mixing chamber of the HAAKE RHEOCORD. In a typical procedure the reactive composition is milled at 170°C with a mixing speed of 64 rpm. Temperature stabilization is reached after 3 minutes of mixing. The reaction is stopped after various times (between 3 and 100 minutes) and the modified mixture cooled before analysis.

Extrusion

The continuous process carried on in an extruder needs an efficient equipment and we selected the twin screw from Werner Pfleiderer (ZSK 30). It is an intermeshing, co-rotating, self wiping twin-screw, characterized by a building block system, allowing various screw profiles, adapted to efficient mixing and adjustable residence time. Figure 1 is a schematic representation of the equipment and its most common screw profile and associated feed systems.

2.3. Analysis :

The modified EVA was washed to remove remaining reaction products and catalyst and purified as indicated in a previous paper (4). To determine the VA (vinyl-acetate) content in the modified copolymers, we used IR spectrocopy, which is the most convenient method, as mentioned before by KOOPMANS (5,1982). But in order to ensure the accuracy of the calibration curve, we measured also the VA content of some copolymers by proton NMR.

DSC measurements were also made on the modified copolymers for the evaluation of crystallinity changes after conversion.

HPLC analysis was used for composition analysis of the reagents extracted by devolatilization.

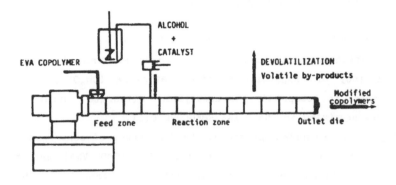

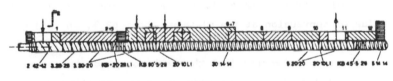

SCREW PROFILE

"Figure l., Werner Pfleiderer twin-screw extruder and associated
feed systems".

3. RESULTS AND DISCUSSION :

3.1. Comparison of catalyst efficiency :

Many compounds are known to act as catalysts for transesterification reactions (6), and we tried first to compare the efficiency of respectively sodium methoxyde (CH_3ONa) and an organo metallic compound : dibutyltin dilaurate (DBTDL) by discontinuous mixing in the HAAKE RHEOCORD cell.

On figure 2, we have represented the dependance of conversion on reaction time in a mixture containing EVA (1040 VNA) and the stoïchiometric amount of primary alcohol, (1-octanol) for both catalysts with the same concentration) (3 p.h.r.). The upper curve (a) shows a maximum of conversion at about 72 % which is obtained after a very short time of reaction, about ten minutes, in presence of sodium methoxyde. The second curve (b) is related to the mixture containing DBTDL and depicts a slower speed of reaction and also a lower value of conversion at the plateau, about 62 %. In this case, most of the reaction has already occured after 30 minutes.

It was also easy to verify that the reaction products contain in both cases octyl-acetate, the main product resulting from the transesterification reaction. In our previous papers (4,6), we have shown that the difference in final conversion values is a consequence of the contribution of saponification resulting from the hydrolysis of sodium methoxyde.

As conversion and speed of reaction seem to be quite high and despite to the side-reaction of saponification, we tested the system containing sodium methoxyde in the twin-screw extruder. The following experimental conditions were used : screw speed, about 150 rpm ; temperature of molten EVA : 170°C and average residence time : 80 sec. In figure 3, we represent the dependence of conversion against the amount of catalyst for various molar ratios. These curves show that a minimum content of about 1.2 phr CH_3ONa is needed to reach the maximum of conversion under the given extruding conditions. As expected from the experiments based on molar ratio changing, it appears that we attain a final value of conversion at

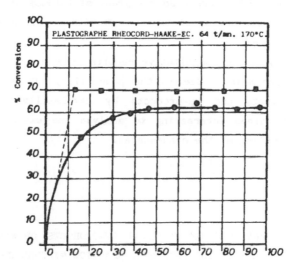

"Figure 2., Dependence of the conversion on the reaction time for
two different catalysts"
● EVA 1040/Octanol/DBTDL: 100/22/3 phr.
■ EVA 1040/Octanol/CH_3ONa: 100/21,3/3 phr.

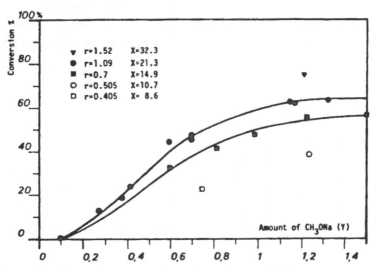

"Figure 3., Reactive extrusion: dependence of the conversion on the
amount of CH_3ONa and on the molar ratio".

equilibrium, characteristic of a reversible transesterification reaction. This study shows also the great efficiency of this system as we obtain a conversion of about 63 % with an average residence time of only 80 sec.

Table 2 reports the quantitative results based on these reactive extrusion experiments. Due to the possibility of recovering the volatile reagents and by-products by devolatilization in the extruder, it seems interesting to analyse the extracted mixture obtained by cooling of the condensates.

Table 2 :   ANALYSIS OF REACTION BY-PRODUCTS OBTAINED IN THE DEVOLATILIZING ZONE OF THE EXTRUDER. REACTION MIXTURE COMPOSITION : EVA 1040 VN4/1-Octanol/CH_3ONa.

Molar ratio Ac/OH	Catalyst (phr)	Conversion (%)	HPLC Analysis	
			octylacetate(%)	octanol(%)
0.98	0.27	13.6	12	88
0.94	0.41	23.2	21	79
0.95	0.59	33	32	68
0.96	0.81	42	40	60
0.95	0.98	48	48	52
0.94	1.16	62.4	63.5	36.5

These analysis were made by high pressure liquid chromatography
(HPLC) and the corresponding results are given in the two last
columns of table 2. The exact balance between octyl-acetate and
octanol in the condensate confirms definitively transesterification
equilibrium

$$EVA + ROH \rightleftharpoons RO\underset{\underset{O}{\overset{\|}{}}}{C} -CH_3 + EVAL.$$

3.2. Reactivity of various alcohols :

- Primary aliphatic alcohols : In order to verify the influence of
reagent solubility in the molten EVA on conversion, we tested a
series of paraffinic alcohols with various carbon chain length, all
used in stoichiometric amounts with respect to VA content of the
EVA copolymer (1040 VN 4). Corresponding results are reported on
table 3.

Table 3 : **REACTIVITY OF ALIPHATIC MONOALCOHOLS**

Nature of alcohol	Alcohol amount (phr)	Conversion (%)
Methanol (C1)	5.3	55
1-Butanol (C4)	12.1	62
1-Octanol (C8)	21.2	72
1-Dodecanol (C12)	31.1	72
1-Hexadecanol (C16)	39.5	69

The experiments carried out in the Rheocord mixing cell show that
the observed decrease in conversion is partly due to the volatility
of the low molecular weight alcohols and therefore maximum with
methanol.

When the length of the hydrocarbon chain increases, it seems that we obtain a miscible or compatible system leading to a maximum conversion of about 70 % for the stoechiometric mixtures.

It is also important to notice that paraffinic alcohols of very high chain length are not of interest for modification by transesterification. The amount needed for high conversion level are too large and the final modified copolymer contains a lot of residual reagents and esters which are difficult to extract by devolatilization.

- <u>Diols and polyols</u> : It seems more interesting to introduce diols and polyols of lower molecular weight and to verify if the primary (or secondary) alcohol groups are able to react independantly in these conditions. Table 4 gives an abstract of the main results obtained with these products, when the reactive mixture is based on the stoichimometric ratio (alcohol/ester) in presence of 3 phr sodium methoxyde, as catalyst. Temperature and duration were fixed, respectively at 170°C and 50 mn. Frequently this large mixing time was not necessary, as can be seen in figure 4 which describes conversion dependance on reaction time with hexanediol-1,6 as reagent.

Both figure 4 and table 4 confirm the high reactivity of a mixture based on hexanediol-1,6, for which we obtain conversion of more than 75 %. Dodecanediol-1,12 leads to the same efficiency and confirms the global reactivity of the two primary alcohols functions. With the low molecular weight products, we notice also a significant decrease in conversion as seen previously for the corresponding monoalcohols.

We tried to understand why ethanediol and butanediol present this unexpected behaviour and compared the effective conversion of ester groups to the resulting secondary alcohols on the modified copolymer.

Table 4 : **REACTIVITY OF DIOLS AND POLYOLS**

Nature of polyol	Amount (phr)	Conversion (%)
1,2-Ethanediol	5.3	52.5
1,4-Butanediol	7.3	67.0
1,6-Hexanediol	9.6	75.5
1,12-Dodecanediol	16.6	75.5
Sorbitol	5.0	12.0

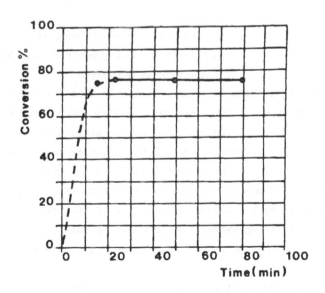

"Figure 4., Reactivity of 1,6 - hexanediol : dependence of the conversion on the reaction time"

Table 5 :   REACTIVITY OF LOW MOLECULAR WEIGHT ALCOHOLS

Nature of the reagent :	Composition (phr)		Conversion (%)	
	Alcohol	Catalyst	Reacted esters	Appeared sec.alcoh.
CH_3OH	5.44	3	76	55
CH_3OH	5.33	1	43	32.5
1-Butanol	12	3.2	87.5	62
1-Butanol	12	1	68	56

Quantitative results reported in table 5 indicate that the discrepancy between reacted esters and appeared alcohols is sensitive to the amount of added sodium methoxyde.

Beside this, it is also important to notice that the variation of torque against time during reactive mixing is very different between the low molecular weight alcohols or diols and 1-octanol as shown on figure 5. The sharp increase in viscosity in the beginning of mixing has to be connected with the possibility of a crosslinking reaction in the case of methanol and butanol, this fact is confirmed by the limited solubility of the reacted mixture. These results are in agreement with preceeding observations made by RÄTZSCH who proposes a mechanism of the crosslinking reaction (7,1971).

Unfortunately, these interesting results are not confirmed in the case of sorbitol, which yet presents two primary and four secondary alcohol groups. It is easy to show that sorbitol has very poor solubility in molten EVA and this is the main reason of the observed limited reactivity.

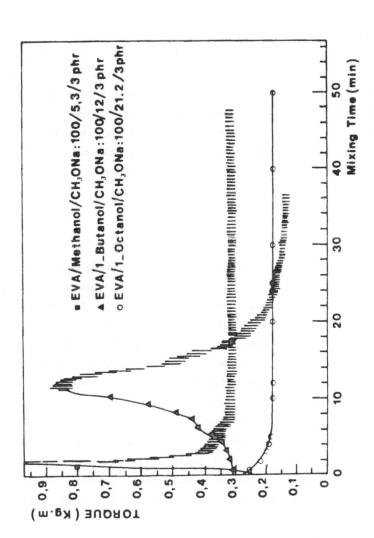

"Figure 5., Discontinuous reactive mixing: dependence of the torque versus reaction time".

3.3. Preliminary conclusions :

At this point of our studies, it appears that the expected transesterification reaction leads to high conversion values, both in discontinuous or continuous mixing conditions. Furthermore, sodium methoxyde appears to be an efficient catalyst but is connected with side reactions, hydrolysis in presence of water and crosslinking with low molecular weight alcohols.Therefore we used preferably for the rest of the work DBTDL, which is an efficient catalyst without any side reaction. In the same time,it is worthwhile to notice that conversion yield is related to the compatibility between reactants and molten copolymer. This last remark and the use of DBTDL would be the guidelines to the study of the kinetics of the transesterification reaction.

4. KINETICS :

Chemical reactions require that at least two molecules or reacting groups come into close proximity so that energy, electrical charge, or chemical groups may be exchanged between them. The process of bimolecular reaction between the ester group fixed on the polymer backbone and the paraffinic alcohol is related to two successive steps of diffusion, bringing the two reactive species in proximity, and chemical reaction successively. This usual presentation of solution behaviour is also applicable to high viscous systems if the molten polymer and the added reagents form an homogeneous mixture. According to preceeding remarks about the correlation between conversion yield and reactant compatibility, we felt necessary to verify by diffusion measurements the true solubility of some alcohol reagents used for our transesterification reaction.

4.1. Diffusion analysis :

These measurements were performed using the technique and experimental conditions described by MOISAN (7,1979). Figure 6 is a

schematic representation of the experimental cell which brings in
close contact, under pressure, a file of 15 EVA thin films (100 um)
and a source containing a large excess (30 %) of the chosen
reactant. The whole system is introduced in an oven at 80° for
about 100 min. The ideal diffusion profile is indicated on figure 7
by the black line going through the films. This signifies that the
concentration at the interface between source and film is constant
and equal to the reagent solubility, S (8,9). However it is
difficult to use this extrapolation for the evaluation of diffusion
constant D or solubility S for Fick's laws calculations. The dotted
line represented in figure 6 is much more representative as
confirmed by the experimental results determined from IR analysis
on each film. We verified also that D is really independent of the
concentration and that the values of D and S deduced from two
experiments based on different diffusion times are identical. The
most important results are given on table 6.

Table 6 : DIFFUSION AND SOLUBILITY DATA OF ALCOHOL REAGENTS IN
           EVA COPOLYMER (80°)

Nature of reagent	Diffusion constant D (cm /s)	Solubility S (% b.w.)	Stoichiom. amount (% b.w.)
1-octanol	5.6 x 10	15.8	17.6
1-dodecanol	4.8 x 10	15.5	23.7
1-hexadecanol	3.6 x 10	14.5	28.3
1,6-hexanediol	-	6 (estim.)	8.7
Sorbitol	-	0	4.8

As indicated in this table, the higher molecular weight primary
alcohols have a significant solubility in EVA at 80°C. Furthermore
it is worthwhile to notice that these values are of the same
magnitude as the amount needed for a stoichometric reactive mixture
used in transesterification reaction. Measurements made   with

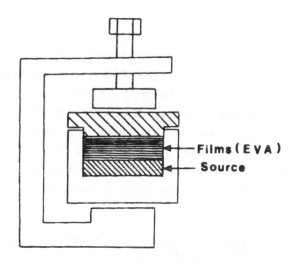

"Figure 6., Experimental cell for diffusion measurements"

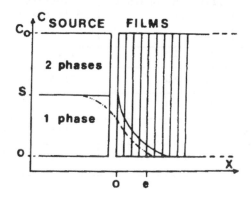

"Figure 7., Schematic representation of theoretical (black curve)
           and   experimental (dotted line) concentration (C)
           profiles versus depth of penetration (x)"

hexanediol-1,6 were difficult to perform because, even for very long diffusion times, the obtained concentration of diol are only detectable in the first three films, leading to the estimated value of solubility. Finally, it was easy to confirm the insolubility of sorbitol in EVA, which is dispersed as solid aggregates in the film, as shown by microscopic analysis on heating plate.

Following MOISAN et BILLINGNAM'S conclusions (8,9,10), the solubility at 170°C will be greater than at 80°C. Thus the reactive mixtures containing paraffinic alcohols like octanol-1, dodecanol-1, and probably hexanediol-1,6 are homogeneous. From this conclusion we carried out kinetical studies for the most simple case, using DBTDL as catalyst.

4.2. Kinetic studies :

The general reaction schema of the equilibrated transesterification reaction is given below :

(1)

$t=0$	a	b	0	0
$t < t_{eq}$	$(a-x)$	$(b-x)$	x	x
$t = t_{eq}$	$(a-\alpha)$	$(b-\alpha)$	α	α

The simplest case occurs when both forward and reverse reactions are of the same order and the formed products are initially absent from the reacting mixture. If both reactions are second-order, according to equation (1) the net rate of disappearance of A is :

(2) $-d\ NA/dt = k_1\ C_A \cdot C_B - k_2\ C_C \cdot C_D$

At the equilibrium, in the case of reversible reactions in ideal solution, $dNA/dt = 0$, then :

(3) $k_1 \cdot C_A \cdot C_B = k_2 \cdot C_C \cdot C_D$ and

(4) $C_C \cdot C_D\ /\ C_A \cdot C_B = k_1/k_2 = K = \alpha^2\ /\ (a-\alpha)(b-\alpha)$

where K is the equilibrium constant and α is the conversion at equilibrium.

Replacing NA, C_A, C_B, C_C, C_D by their values, we get :

$$dx/dt = k_1\ (a-x)\ (b-x) - k_2\ x^2$$

(5) $dx/(x^2 - (a+b)mx + mab) = (k_1 - k_2)\ dt$

with : $m = k_1\ /\ (k_1 - k_2)$

from equation (4) and the expression resulting from integration of equation (5), both speed constants can be calculated :

(6) $1/X_1 - X_2\ \ Ln\ X_2\ (X_1 - x)\ /\ X_1\ (X_2 - x) = (k_1 - k_2)t$

X_1 and X_2 are the roots of : $x^2 - (a+b)\ mx + mab$ (cf equation (5))

From the knowledge of the equilibrium constant and graphical representation of equation (6), it is possible to evaluate the two kinetic constants k_1 and k_2. But we found it convenient to use a numerical solution of equation (6). From initial and experimental concentrations, k_1 and k_2 are adjusted in order to fit calculated and experimental conversion values. This program does not need the prior evaluation of the equilibrium constant.(5)

Figure 8 shows the calculated variation of conversion (expressed in mole/liter) in function of time for three different temperatures, respectively 150°C, 170°C and 190°C. the smooth curves calculated with the first experimental data (t= 40 mn) fit all the experimental results up to 100 minutes. Table 7 reports the corresponding quantitative values of k_1, k_2 and K, calculated value of equilibrium constant which is compared to the experimental one. From equation (5) and experimental error on conversion, we are able to the estimate corresponding uncertainty in k_1 and k_2, confirming the good agreement between the computed and experimental equilibrium constants.

Besides this, it seems interesting to compare kinetical calculations and experimental data for various alcohols and for different EVA copolymers with a VA content between 5 and 28%.

All these experiments were carried out, using the discontinuous mixing device and DBTDL as catalyst at 170°C. Figure 9 shows clearly that if solubility is an important parameter, chemical reactivity depends also on the nature of the alcohol. Secondary alcohol function of octanol-2 leads to poor efficiency in conversion (32 % at equilibrium), whereas octanol-1 and hexanediol-1,6 present the same high values of about 62 %. In the case of the two triols : 1,2,6 - hexanetriol and trimethylolpropane, limited conversion is due to both decreased reactivity and solubility of the two products and to steric hindrance in the second case.

Table 8 presents the results obtained with different EVA copolymers and three compositions. Conversion values at equilibrium are identical for the whole composition range with the low VA content products (5 % and 14 % b.w.). but the last product, containing 28 % b.w. VA has a higher transesterification efficiency. It is well known that at high percentage of VA content, the random distribution of VA units is not respected. Taking into account the formation of diads and triads, the enhancement of conversion could be attributed to proximity catalytic effects.

Table 7 :   COMPUTED RATE CONSTANTS AND CORRESPONDING INCERTITUDE

Temperature (°C)	$k_1 \times 10^2$ $(1.m^{-1}.mn^{-1})$	$k_2 \times 10^2$	K	K_{exp}
150	2.0 + 0.2	0.66 + 0.05	3.1	-
170	5.3 + 0.4	2.2 + 0.2	2.4	2.4
190	13 + 0.8	3.1 + 0.5	4.1	3.7

Table 8 :   INFLUENCE OF EVA COMPOSITION ON EQUILIBRIUM CONVERSION (170°C)

E V A (composition)	MOLAR RATIO Ac/OH	CONVERSION (%)
1010 V G 2 (5 % A V)	0.5	39
	1.0	61
	2.0	77.5
1040 VN 40 (14 % A V)	0.5	38
	1.0	62
	2.0	78*
OREVAC 9006 (28 % A V)	0.5	44
	1.0	68.5
	2.0	-

* extrapolated from calculated curves.

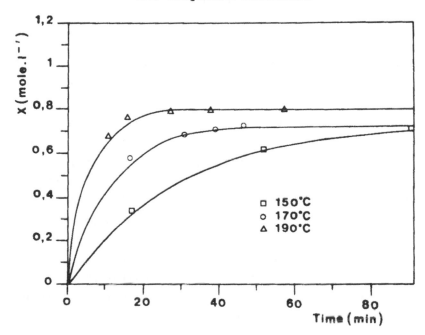

."Figure 8., Influence of temperature on conversion versus reaction time".

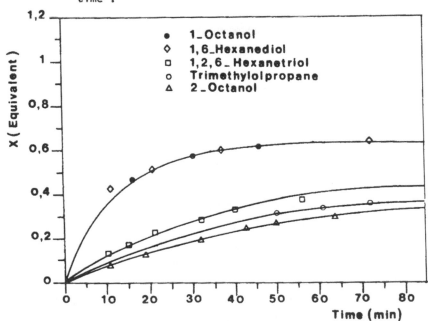

"Figure 9., Influence of alcool nature on extent of conversion versus time".

5. CONCLUSION :

The transesterification reaction of molten EVA, in presence of paraffinic alcohols and basic catalysts, leads to high conversion of the ester groups to secondary alcohol in both discontinuous and continuous processing equipments. Various kinds of alcohols and two different catalysts have been used. Sodium methoxyde is a powerful catalyst for the equilibrated transesterification reaction but we observed also side reactions, like crosslinking with low molecular weight alcohols and hydrolysis of the catalyst followed by partial saponification of the EVA. Kinetical studies were performed in presence of DBTDL, which is an efficient catalyst without any side reactions. The solubility of the main alcohol reagents were verified by diffusion measurements. The general reaction scheme and the related kinetics, corresponding to an homogeneous system, leads to a fair evaluation of the rate constants.

ACKNOWLEDGEMENT

The authors gratefully acknowledge the support for this research project from the ATOCHEM Company.

REFERENCES

1. Jap. Patent N° 72 - U30932 (1972)
2. RDA Patent N° 107.938 (1974)
3. US   Patent N° 3.972.865 (1976)
4. A. Bouilloux, J. Druz, M. Lambla (1986)
   Polym. Process Eng. : Under press
5. R.J. Koopmans, R. Van der Linden and E.F. Vansant (1982)
   Polym. Eng. & Sci., 22, n° 14

6. A. Bouilloux, PhD dissertation (1985) Université de Strasbourg

7. H. Rätzsch & U. Hofman (1971)

   J. Appl. Polym. Sci., 15, 589-595

8. J.Y. Moisan, PhD dissertation (1979) Université de Rennes

9. J.Y. Moisan (1983) Europ. Polym. J. 16, 1127

10. N.C. Billingham & P.D. Calvert (1981)

   J. Appl. Polym. Sci., 26, 3543.

REACTIVE BLENDING: STRUCTURE AND PROPERTIES OF CROSS-LINKED OLEFINIC THERMOPLASTIC ELASTOMERS.

D. Romanini, E. Garagnani and E. Marchetti

HIMONT ITALIA S.p.A.-Centro Ricerche Giulio Natta,

44100 FERRARA (Italy)

ABSTRACT

Thermoplastic elastomers are characterized by the presence of thermoreversible interactions among the polymeric chains.

The crosslinked olefinic thermoplastic elastomers (CTPO) are processable by the same technologies used for the thermoplastics and show good elastic properties.

The morphologic structure in which the crosslinked rubber particles appear homogeneously and finely dispersed within the polypropylenic matrix has been investigated.

1. INTRODUCTION

Thermoplastic Elastomers (TPE) are a class of polymers of heterophasic nature and with a low elastic modulus. They are characterized by the presence of thermoreversible interactions among the polymeric chain. As a result of their particular molecular configuration, thermoplastic rubbers can be processed in the molten state by using the same technologies used for thermoplastic materials. However, at working temperatures, the elastic-mechanical properties of the manufactured items are essentially the same as simi-

lar items based on conventional vulcanized elastomers. In
most thermoplastic rubbers of industrial interest, low
heat resistance bonds have been achieved by the use of
block copolymers and, more recently, by mechanical blending
of elastomers and plastomers.

The thermoplastic elastomers commercially available today
can be classified into the following main classes:

STYRENE BLOCK COPOLYMERS - Obtained by anionic sequential
polymerization. The following pertain to this class:

a) both the linear and radial three-block polymers, i.e.
   styrene-butadiene-styrene (SBS);

b) the linear three-block polymers, i.e. styrene-isoprene-
   styrene (SIS); and

c) the block polymers, i.e. styrene-ethylene-butene-styrene
   (SEBS).

POLYETHER-ESTER THERMOPLASTIC RUBBERS (COPE) - Obtained by
polycondensation.

POLYURETHANE THERMOPLASTIC RUBBERS (TPU) - Obtained by reaction between a diisocianate and a polyol in the presence
of a chain extender.

OLEFINIC THERMOPLASTIC RUBBERS (TPO) - Obtained by the physical blending of a plastic material, generally polypropylene (PP), with an EPM or EPDM elastomer. After blending,
a double-phase structure is obtained with the greatest component (rubber) representing the continuous phase where
the minor component is fully immersed in the form of small
particles.

CROSS-LINKED OLEFINIC THERMOPLASTIC RUBBERS (CTPO) - Obtained by dynamic vulcanization of the elastomeric phase
during blending with the plastomer (reactive blending).

This work deals with this last class of thermoplastic elastomers which appear to be of great interest thanks to a
number of features, such as:

- preparation technology, which utilizes compounding machines as "reaction chambers";
- the resulting morpologic structure, for which the minor
  component is the continuous phase;
- the elastic-mechanical properties nearly approaching those of a vulcanized rubber.

2. STRUCTURAL MODEL

To obtain thermoplastic rubbers processable with the
typical thermoplastic technologies and exhibiting good
elastic properties, it is necessary to follow a particular
structural model. It requires PP as the continuous phase
and spherical particles of completely crosslinked rubber
as the discrete phase.

The accomplishment of such a model, therefore, requires
the achievement of a completely crosslinked dispersed phase. Due to its low deformability, this phase behaves as a
filler. However, because of its chemical nature, it is
capable of exhibiting the necessary adhesion. The latter
feature, allows for a good stress transfer between the
thermoplastic matrix and the elastomeric phase, enabling
the CTPO to exhibit excellent elastic-mechanical characteristics. The low deformability of the dispersed phase allows the rubber particles to maintain a spherical form and
prevents the aggregation phenomena improving the product
processability.

This system presents the lowest value of effective dispersed phase fraction, $\emptyset$eff, and is thus characterized by a

minimum viscosity. Relative viscosity, ηr, is correlated
to $\emptyset$eff, according to the following equation:

$$\eta_r = (1 + 1.25 \emptyset eff)^2 \qquad (1)$$

where: $\qquad \emptyset eff = \dfrac{Vp}{Vt - Vv} \qquad (2)$

Vp = particle volume

Vt = total volume

Vv = volume of the aggregates including the interstitial
vacuums.

These interstitial vacuums can be filled by the PP molecules which, however, do not contribute to the system flo-
wability because they remain unmovable /⁻1_7.

The equation above reported can be rewritten also as follows:

$$\emptyset eff = \dfrac{\emptyset}{(1 - \emptyset/\emptyset m)} \qquad (3)$$

where $\emptyset$ and $\emptyset m$ represent the volumetric fraction and the
maximum packing fraction of the dispersed phase, respecti-
vely. $\emptyset m$ decreases either with increase in the particle
shape ratio or due to aggregation phenomena. Thus $\emptyset eff$
increases and brings about an increase in the system viscosity /2̄,3_7.

3. PREPARATION TECHNOLOGIES (REACTIVE BLENDING)

In order to obtain a product with the structure before
specified, two well defined sequences must be followed:
1) homogeneization and dispersion of the EPDM into the PP
2) dynamic crosslinking of the elastomer.
During the first step in the traditional compounding machines, the material crushing continues until the stresses
on the particles equalize the interface tension.

To this purpose, Taylor worked out a mathematical treatment
$\underline{/}$ 4 $\underline{/}$, demonstrating that the best dispersion is obtained
when:

- interface tension is low (good adhesion);
- the dispersed phase viscosity is low in comparison to
  that of the matrix;
- the shear stresses applied in the blending phase are high.

The second step, which provides crosslinking of the elastomeric phase in the presence of a plastomer and high frict-
ional stresses, is called "in situ" dynamic vulcanization.
Beyond helping the dispersion of the crosslinking system
in the polymeric mass, the presence of high shear rates
$\dot{\gamma}$, allows the EPDM chains to extend and react more quickly with the crosslinking agent.

The frequency factor A, which appears in the Arrhenius
equation, is higher with the extend macromolecular conformations than with the coiled ones $\underline{/}$ 5 $\underline{/}$.

In fact, the extended chains have higher average quadratic
gyration radii and, greater collision diameters than do
coiled macromolecules.

The increased reaction rate reduces the material residence
time in the compounding equipment. Beyond increasing the
hourly production capacity of CTPO, this restricts degradation phenomena of the polyolefines subjected to high
temperatures. Consequently, mechanical properties are
improved.

However, it must be pointed out how high shear rates may
even negatively affect the elastic properties of CTPO, particularly when the elastomeric phase is not crosslinked
enough.

This is due to the fact that since elasticity is dependant
on a variation in entropy $\underline{/}$ 6 $\underline{/}$, it decreases upon an increase in the number of extended macromolecular conforma-
tions formed in CTPO during the dynamic vulcanization process.

It is to be born in mind, however, that the relaxation time of such extended conformations decreases with an increase in their established crosslinking density.

Blends consisting of 60% EPDM and 40% PP, vulcanized dynamically in a double-screw extruder at 230°C with a crosslinking system based on phenolformaldehyde resin, exhibit
different tensile properties as a function of the shear
rate imposed during this "in situ" dynamic vulcanization
process (figure 1).

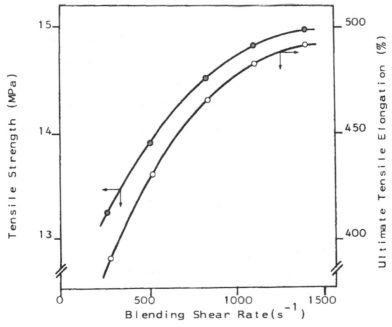

Fig.1 – Tensile strength and ultimate tensile elonga-
       tion versus blending shear rate $\underline{/}\overline{7}\underline{/}$ of EPDM/PP
       (60/40 wt.) dynamically crosslinked.

It is necessary to recall, however, that, during the vulcanization phase, the presence of high shear rates brings
about a better dispersion of the crosslinked elastomeric
phase as well as a reduction in the dimension of the discrete domains.

This causes the shift of the pseudoplastic flow limit towards higher shear rate values in the flow curve $\eta/\dot{\gamma}$ /8/.

4. STRUCTURE

4.1.Molecular structure

The CTPO molecular structure is characterized by the
crosslink density of the elastomeric phase. There are various techniques able to evaluate the crosslink density
of a polymeric material /9/.

Among the methods tested for determining the average molecular weight Mc of CTPO elastomeric phase between two
crosslinks, two of them give reliable results.

The first method requires the determination of the Young
modulus, E, by mechanical procedures and the use of the
formula /6/:

$$E = \frac{3 \, d \, R \, T}{Mc} \qquad (4)$$

where: T = absolute temperature in °K

R = gas constant

d = polymer density

The second method requires the determination of the Young
modulus on swelled specimens in a suitable solvent and by
the use of the formula suggested by Cluff and coworkers
/10/:

$$\frac{d}{Mc} = \frac{ho \, S \, go}{3 \, A \, R \, T} \qquad (5)$$

where: ho = thickness of the unswelled and undeformed specimen,

A = cross-sectional area of the unswelled and undeformed specimen,

S = slope in g cm of the straight load/deforma-
tion line,

go = gravity acceleration.

Although more laborious than the first, this second method
proves to be more reliable and reproducible.

On some CTPO consisting of 60% EPDM and 40% PP and vulcanized dynamically with a phenolformaldehyde resin, xylene
extractions were performed in a vapour-jacketed Soxhlet
apparatus in order to remove the PP resin that can interfer
with the rubber swelling process due to the solvent /11/.
Crosslinked rubbers isolated from PP were subjected to
the Cluff method for the determination of the crosslink
density. EPDM rubbers, statically crosslinked without PP,
were likewise subjected to the same method.

From the results reported in Fig.2 it can be realized that
the rubber crosslink density is slightly affected by the
presence of PP. The crosslink density may be a parameter
able to select CTPO with completely xylene-insoluble elastomeric phases. In fact, over a given crosslink density,
the elastomeric phase becomes insoluble in any solvent.

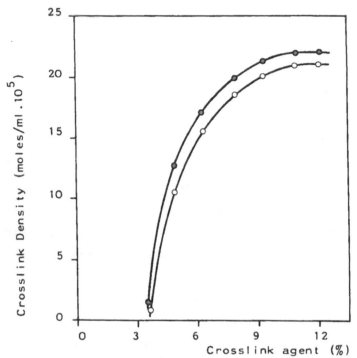

Fig.2 – Crosslink density of EPDM rubber in presence
of PP(o) and without PP (•) measured according
to Cluff method.

4.2. Morphological structure

Fig. 3 shows the morphology model of PP/EPR blends as
a function of composition and melt viscosity of the components. Though always occurring near the intermediate compositions, the phase inversion is a function of the melt
viscosity ratios of the two components. For ratios approaching 1, the zone where the phase inversion takes place is
broad and characterized by the presence of co-continuous
phases within a wide composition range.

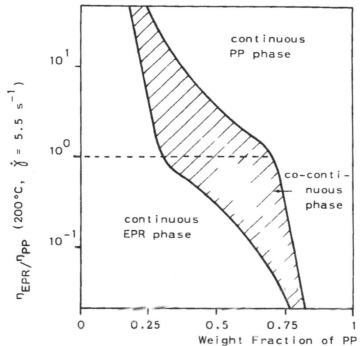

Fig.3 - Morphology model of EPR/PP blends.

If these ratios move away from the unit, the zone where
the phase inversion occurs becomes narrow. It moves towards
more EPR-or PP enriched phases depending on whether the
η EPR/η PP ratio is greater or lower than 1 /̅12, 13̅.
Due to high viscosity of the completely crosslinked rubber,
PP represents the continuous phase in CTPO even when present in very small concentrations.
The above was verified by means of a morphological study
of uncrosslinked (TPO) and crosslinked (CTPO) PP EPR blends
prepared with rubbers properly modified with inorganic
tracing elements such as Al and K. By using combined electronic microscopy and energy dispersive microanalysis
(EDS) techniques /̅14̅7, it was, in fact, possible to

identify the rubber location both in TPO and CTPO.

Upon scanning electronic microscopy (SEM), the fracture
surface of a TPO consisting of 20% rubber modified with
Al, and 80 PP, presents spherical inclusions immersed in
a plastomeric matrix (figure 4). EDS microanalysis (fig.5)
carried out on dispersed rubber particles (spectrum A) indicates the presence of Al, while the spectrum B, obtained
under the same conditions on the matrix, does not. A direct
comparison of the two spectra, shows that the Al bonded to
the rubber is only present in the dispersed particles.

The fracture surface of a CTPO, consisting of 75% oil-ex
tended EP rubber modified with K and crosslinked with
phenolformaldehyde resin and 25% of PP presents, upon SEM
analysis, a dispersed phase consisting of dark spherical
particles ranging in size from 5 to 0.1 μm (figure 6).

The fracture surface looks more PP-enriched than the CTPO
composition due to the fact that the fracture itself, started on a specimen immersed in liquid nitrogen, propagates
in PP.

EDS microanalysis (figure 7) carried out on dark particles
(spectrum A) points out a higher K content than that observed in the matrix (spectrum B).

The greater K content of the discrete phase proves the
rubber is the dispersed phase.

Thus, in CTPO, PP is the continuous phase even though it
is the minor component.

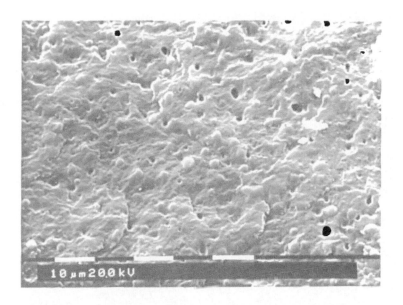

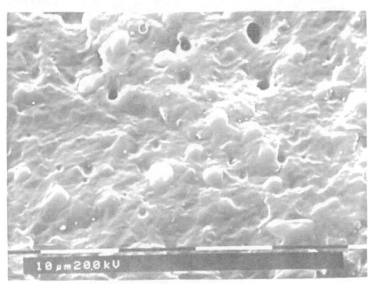

Fig. 4 - SEM micrograph of TPO
(EPDM PP ratio: 20 80 wt.)

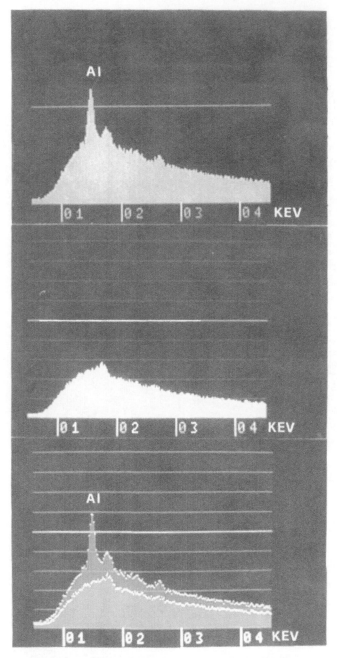

Fig.5 - EDS Spectra of TPO (EPDM/PP ratio: 20/80 wt)

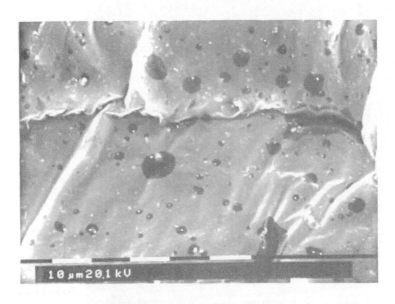

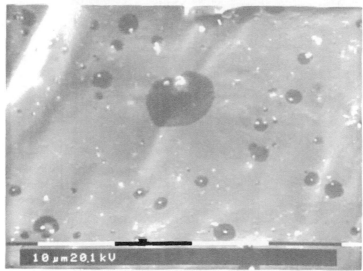

Fig. 6 - SEM micrograph of CTPO
(EPDM/PP ratio: 75 25 wt.)

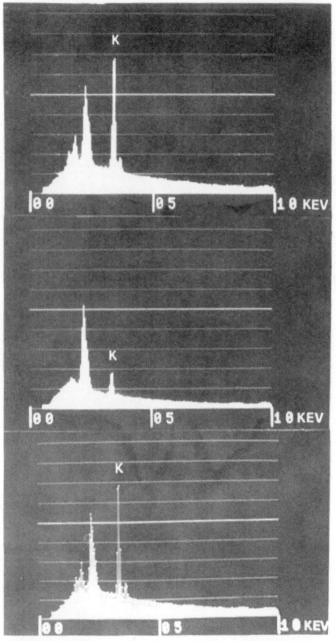

Dispersed
phase
Spectrum A

Continuous
phase
Spectrum B

Overlapping
of Spectra A
and B

Fig.7 - EDS Spectra of CTPO (EPDM/PP ratio: 75/25 wt.)

5. PROPERTIES

5.1.Rheological properties

From the flow curves determined by capillary rheometer reported in figures 8 and 9, it can be seen that the pseudoplastic behaviour (viscosity variation with the shear rate $\dot{\gamma}$) is more evident in CTPO than in TPO. This can be explained by considering that at low shear rates long range molecular relaxation processes prevail, whereas at high shear rates the flow is affected by the short range molecular relaxation processes and thus by low relaxation times according to Rouse theory [15].

In the crosslinked rubbers, the intermolecular junctions prevent long range motion and thus the flow processes involve only short range motions, for low shear rate values already. In fact, while in TPO relaxation times increase with a decrease in the shear rate, in CTPO such times remain practically constant upon a change in shear rate.

From the Pao equation [16]:

$$\eta = \frac{G \; \lambda}{1 + \dot{\gamma}^2 \; \lambda^2} \qquad (6)$$

where: G = shear   dulus

λ = relaxation time.

it can be deduced that with a decrease in the shear rate, viscosities increase more in CTPO than in TPO.

CTPO rheological behaviour is also affected by the size of the completely crosslinked rubber particles. In fact, the viscosity in such systems increases as the size of the rubber particles decreases.

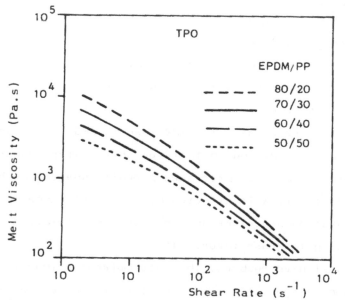

Fig.8 – Flow curves of different composition TPO
by Instron Capillary Rheometer at 230°C.

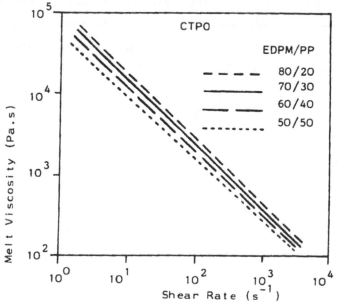

Fig. 9 – Flow curves of different composition CTPO
by Instron Capillary Rheometer at 230°C

This is shown in figure 10 which reports the influence crosslinked rubber particle size has on the extrusion pressure of a CTPO consisting of 70% EPDM and 30% PP.

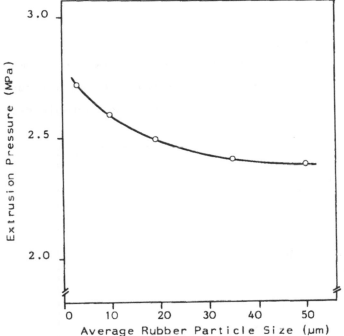

Fig. 10 - Processability of CTPO versus particle size of crosslinked EPDM rubber.

These tests were carried out in a single-screw extruder at a constant output rate of 7 g/min at 230°C.

The determination of the average rubber particle sizes in CTPO was carried out with a method consisting in fracturing the polymeric material in nitrogen liquid and then observing at least 100 cured rubber particle by scanning electronic microscopy.

The behaviour reported in fig.10 suggests an interaction among the crosslinked rubber particles, whose energy increases upon a decrease in the particle size.

Such interactions depend on the surface area of the dispersed particles: by increasing the ratio between particle
surface area and volume the reciprocal interactions become
more intense.

5.2. Mechanical Properties

CTPO, unlike TPO, give good performance in terms of
elasticity and stiffness within a wide temperature range.
In fact, mechanical properties of TPO considerably decrease upon an increase in temperature.

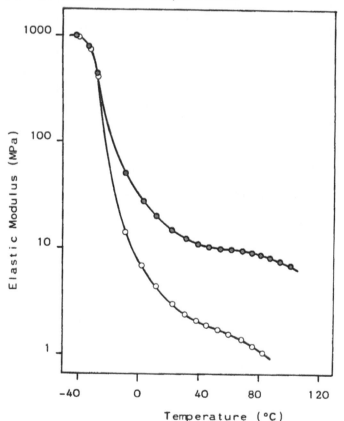

Fig. 11 - Elastic modulus of CTPO (●) and TPO (o)
with 80/20 wt.% EPDM/PP ratio by means
of DMTA.

This can be observed in figure 11 which reports the elastic
modulus of both CTPO and TPO determined as a function of
temperature by means of a dynamic mechanical thermal analyzer (DMTA) with a scanning velocity of 2°C/min and a
frequency of 1 Hz.
The effect of composition, in terms of the PP/EP rubber
ratio, on the main mechanical properties can be noticed in
figures 12 to 16 for both TPO and CTPO.

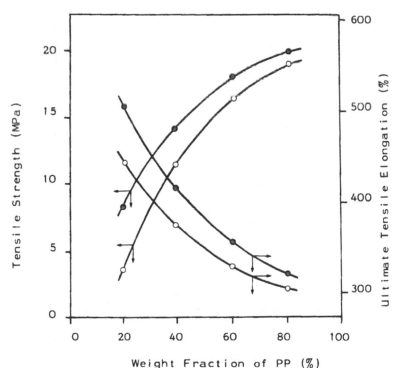

Weight Fraction of PP (%)
Fig. 12 - Tensile strength and ultimate tensile
elongation of CTPO (•) and TPO (o)

Figure 12 shows the higher tensile strenght and elongation at break of CTPO. Also creep resistance is better in CTPO because TPO flow too fast (figure 13).

It is interesting to point out the relationship existing between surface hardness and the polypropylenic fraction of the thermoplastic elastomers (figure 14). The cross-linking of the rubber phase does not affect the surface hardness of the product.

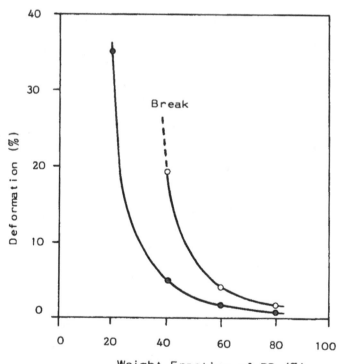

Fig. 13 – Creep in tensile test at 110°C under a load of 0.2 MPa of CTPO (●) and TPO (o)

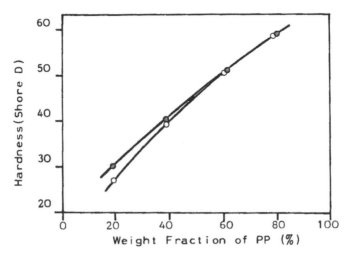

Fig. 14 - The effect of PP content on hardness
of CTPO (●) and TPO (o)

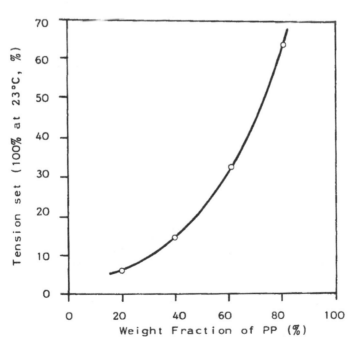

Fig. 15 - The effect of PP content on tension set
of CTPO at 23°C

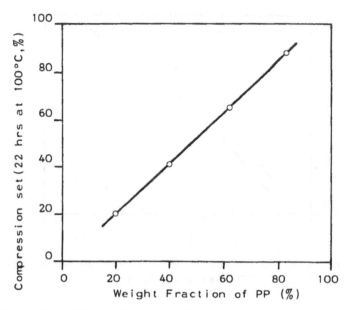

Fig. 16 - The effect of PP content on compression
set of CTPO at 100°C

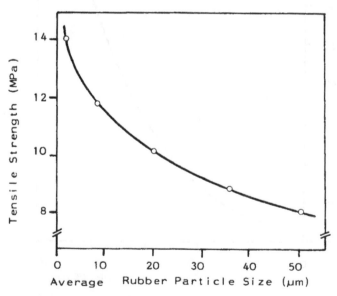

Fig. 17 - Tensile strength versus rubber particle
size of CTPO (60% EPDM, 40% PP)

CTPO tension   t and compression set, mea ured according
to ASTM D-412 and ASTM D-395, ma   worse upon increase in
PP content (figure 15 and 16).
Upon an incr   in the amount of PP, in compari on to the
EP rubber, compositions become less similar to elastomers
and more similar to plastics: hardness, tension set, compression set, tensile strength and creep resistance increase.
Mechanical properties of CTPO, however, depend on the size
of the rubber particle.
Figure 17 reports tensile strength as a function of the
crosslinked rubber particle size of CTPO consisting of 60%
EPDM and 40% PP. From this figure a decrease in tensile
strength upon an increase in the dispersed particle size
can be observed.
Dynamic crosslinking (reactive blending) allows finely
dispersed crosslinked rubber particles to be obtained and
thus provide CTPO with great tensile strength.

5.3. Chemical Properties

      Generally, CTPO exhibit an excellent resistance to
acids, bases, alcohols and glycols: slight resistance to
oils and hydrocarbon fluids and poor resistance to fuels.
In particular, oil resistance is a function of the crosslink   density of the rubber particles $\overline{17}^-$.
Tests of resistance to oil ASTM no.3, carried out on both
TPO and CTPO, showed the trends in figure 18. The volume
swelling percentage (q-1) 100, reported in this figure as
a function of the polypropylenic fraction, has been obtained from the equation:

$$q - 1 = (R/P-1) \ Dc/Do \qquad (7)$$

where:

P and R are the specimen weights before and after swelling
respectively;

Dc is the CTPO density;

Do is the oil density.

As TPO presents an uncrosslinked elastomeric phase, it
shows clearly poorer resistance to oils than that of CTPO.

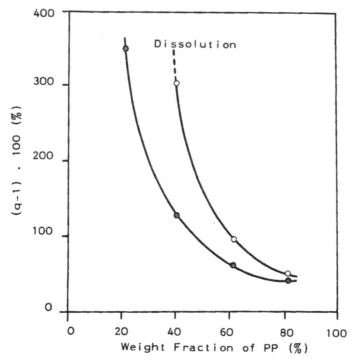

Fig. 18 - Volume swelling, after 166 hrs of immersion
in ASTM N°3 oil at 100°C, of CTPO (•) and
TPO (o)

6. PROCESSING AND MAIN APPLICATIONS

One of the main reasons for the success of TPE is the
simplification they have brought about in the processing
of manufactured items as compared to vulcanized rubbers.

As a matter of fact, TPE usually require only the melting
and molding phases. On the contrary, thermosetting rubbers
require melting, blending, shaping and vulcanization. All
these phases require a high energy consumption and a great
deal of manpower. Furthermore, only a small part of the
scraps can be completely reused.

In terms of mechanical properties, CTPO can surely compete
with the traditional thermosetting rubbers based on EPDM,
poliisoprene (PI and NR), styrene-butadiene (SBR) and poly-
chloroprene (CR) (table 1). Mechanical and elastic properties, resistance to heat, oil and ageing are comparable to
the values offered by these typical thermosetting rubbers.
Likewise, with respect to the other commercially available
thermoplastic elastomers, CTPO produced by the use of the
reactive blending technique, present an optimal performance
/cost balance. In fact, it has been seen that TPO only have
good performance within a restricted temperature range; on
the other hand, some TPE, such as TPU and COPE, offer a
better combination of properties (table 2) but they are
very expensive.

Although no single thermoplastic material nor thermosetting rubber is able to meet all the requirements of the industrial technical items, CTPO makes it possible to introduce thermoplastic rubbers in those fields today covered
by the thermosetting rubbers.

Table 3 lists a group of manufactured items in which CTPOs
are commercially used.

TABLE 1 - CTPO AND COMPETITIVE THERMOSET ELASTOMERS PROPERTIES

	Mechanical properties	Elastic properties	Heat resistance	Oil resistance	Ageing resistance
CTPO	****	****	****	***	****
ISOPRENE THERMOSET	****	***	***	*	***
STYRENE-BUTADIENE THERMOSET	****	****	****	**	***
CHLOROPRENE THERMOSET	****	***	****	****	****
ETHYLENE-PROPYLENE THERMOSET	***	****	****	***	*****

*   Very low
**   Low
***   Fair
**** Good
***** Very good

TABLE 2 - COMPARISON OF THERMOPLASTIC ELASTOMERS PROPERTIES

	Mechanical properties	Elastic properties	Heat resistance	Oil resistance	Ageing resistance
CTPO	****	****	****	***	****
TPO	***	**	***	**	***
STYRENE BLOCK COPOLYMERS	***	*	*	*	**
URETHANE	****	***	***	*****	******
POLYESTER	*****	****	****	******	******

*	Very low
**	Low
***	Fair
****	Good
*****	Very good

TABLE 3 - MAIN CTPO APPLICATIONS

AUTOMOTIVE:

- Car spoilers
- Weather strips (window, windscreen, backlight)
- Gaskets for doors and bonnet
- Vacuum tubing and radiator hoses
- Bellows

ELECTRIC/ELECTRONIC:

- Body plugs
- Switching devices
- Transformer connectors
- Power cable insulation

CONSUMER PRODUCTS:

- Profiles for door and window frames
- Gaskets
- Wheels (caster wheel, roller skates, etc.)
- Toys
- Heels for traditional footwear

7. CONCLUSIONS

During the last ten years thermoplastic rubbers have
expanded greatly, much more so than have thermosetting rubbers. Thanks to the simplicity of the transformation process, TPE are now preferred to vulcanized rubbers in many
fields of application, and their consumption is expected
to continuously increase in the future as well (figure 19).
Though appreciated for the great resistance to ageing and
chemical agents, TPO only cover a limited portion of the
applications of vulcanized rubber due to their poor elasticity at high temperatures.

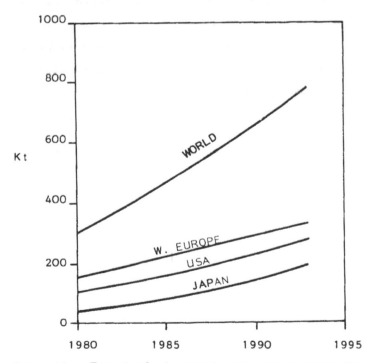

Fig. 19 - Trend of thermoplastic elastomer demand.

By means of a dynamic crosslinking process, it is possible
to obtain a morphological situation where very fine crosslinked rubber particles are homogeneously dispersed within
a thermoplastic matrix of polypropylenic nature.

By properly controlling the particle size, the degree of
rubber crosslinking and the percentage of polypropylene,
it is possible to obtain materials with different hardness
showing excellent elastomechanical properties within a wide temperature range, and, at the same time, the typical
processability of the thermoplastic materials.

Thanks to their properties, CTPO have found many applications and have covered the gap existing between the low
cost traditional thermoplastic elastomers (polyolefinic
and polystyrenic), the traditional vulcanized rubbers and
the high performance, high cost, thermoplastic elastomers
(polyurethanes and copolyesters).

REFERENCES

/ 1 / Fedors R.F., Polymer 20, 324 (1979)

/ 2 / Ferrigno T.H., SPE 34th Annual Technical Conference
          pp.606-608 (Atlantic City, April 26-29, 1976).

/ 3 / Ferrigno T.H., Polymer Engineering and Science 18,
          33 (1978).

/ 4 / Taylor G.J., Proc. Roy. Soc.(A) 138, 41 (1932).

/ 5 / Simonetta M.,"Chimica Fisica",vol.1,part 2,chapter
          2, pp. 247-295, Manfredi Editore (Milano 1966).

/ 6 / Treloar L.R.G.,"The Physics of Rubber Elasticity",
          chapter 3, pp. 40-55, Oxford University Press
          (Londra 1949).

/ 7 / Rodriguez F., "Principles of Polymer Systems", chap-
          ter 7, pp. 155-198, Mc Graw-Hill (Singapore 1985).

/ 8 / Chapman F.M., Lee T.S., SPE 27th Annual Technical
          Conference, pp. 293-298 (Chicago, May 5-8, 1969).

/ 9 / Harrison D.J.P., Yates W.R., JMS-Rev. Macromol. Chem.
          Phys. C 25 (4) 481 (1985).

/10 / Cluff E.F., Gladding E.K., Pariser R , Journal Poly-
          mer Science 45, 341 (1960).

/11 / Coran A.Y., Patel R., Rubber Chemistry and Technolo-
          gy 53, 141 (1980).

/12 / Garagnani E., Romanini D., Moro A., "Giornate di
          Studio su Sistemi Polimerici a due Componenti",(Pisa,
          27-28 Settembre 1984).

/13 / Danesi S., Porter R.S., Polymer 19, 448 (1978).

 14 / Chandler J.A., Audrey E., Glauert M., "X-Ray Micro-
          analysis in the Electron Microscope", chapter 5,
          pp. 425-470, Elsevier North-Holland (Amsterdam 1977)

 15 / Rouse P.E., Jour. Chem. Phys. 21, 1272 (1953).

 16   Pao Y.H., J. Appl. Phys. 28, 591 (1957).

 17 7 Coran A.Y., Patel R., Williams D., Rubber Chemistry
          and Technology 55, 1063 (1982).

SIN INTERPENETRATING POLYMER NETWORKS-STRUCTURE, MORPHOLOGY AND PHYSICAL PROPERTIES

H.L. Frisch
Department of Chemistry
State University of New York at Albany
Albany, N.Y. 12222 U.S.A.

ABSTRACT

Interpenetrating polymer networks (IPN's) are relatively novel combinations of two or more crosslinked polymer networks held together primarily by permanent entanglements (catenation) rather than mutual covalent bonds (grafting). IPN's can be formed by sequential or simultaneous crosslinking of the component networks; in the latter case the resulting IPN's are designated as SIN IPN's. For weakly crosslinked rubbery networks with positive interaction parameter between the networks theory suggests that no sequentially formed IPN would be stable (to micro phase separation) over all network compositions but stable phases could occur for all IPN's and pseudo-IPN's (in which one network is not corsslinked). Many investigations of SIN-IPN's have appeared in over more than a decade, carried out by a variety of groups, which we shall briefly summarize. Our principal focus though will be on the compatible crosslinked poly(2,6-dimethyl-1,4-phenylene oxide) (PPO)-poly-styrene (PS) and related IPN's, the incompatible polyurethane - poly(methacrylate), polyurethane - epoxy and the three component, polyurethane - epoxy - poly(methacrylate) IPN's and the stiff polydiacetylene and polydiacetylene-epoxy IPN's studied recently by us. These studies included determination of the stress-strain curve and ultimate mechanical properties according to standard ASTM procedures. Dynamical mechanical properties were studied with a rheovibron. Glass temperatures were also obtained from differential scanning calorimetry. Micrographs were obtained both by transmission and scanning electron microscopy.
The PPO-PS IPN's were single phase materials with a single glass transition temperature (Tg). The tensile strength to break curve as a function of PPO weight percent showed a well-defined maximum at about 75%. The PU-polymethacrylate IPN's and PU-epoxy IPN's exhibited a variety of morphologies and properties dependent on the type of polymer (e.g. NCO/OH ratio), molecular weight of precursors, presence of charge groups of either sign and the presence of intentional grafts between the component polymer networks. In general, decreasing molecular weight of prepolymers, presence of intentional grafts and/or charge groups of opposite charge results in enhanced homogeneity (due presumably to more interpenetration) and this in turn results in enhanced mechanical properties. Both single and two phased compositions of poly-diacetylene - epoxy and two different polydiacetylenes could be

prepared. These systems exhibited a number of unusual properties including Tg's larger than those of either pure component.

INTRODUCTION

Interpenetrating polymer networks (IPN'S) (See Figure 1) are novel blends of crosslinked polymers held together by predominantly permanent entanglements (catenation). As such they are a special case of macromolecular topological isomerism[1]. The latter has been observed in macrocycles either naturally

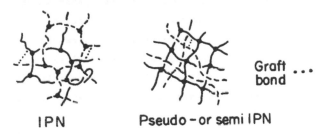

Graft ...
bond

IPN   Pseudo – or semi IPN

Figure 1. Schematic diagrams of IPN's and pseudo (semi) IPN.

as in the circular DNA's or has been synthesized by chemical or biochemical methods producing catenanes, knots as well as more complicated isomers[2] (See Figure 2). In the case of the catenated circular DNA's the individual molecules can be resolved

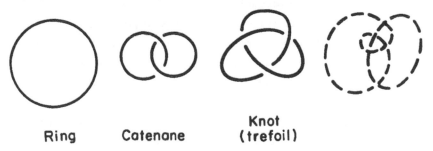

Ring   Catenane   Knot
(trefoil)

Figure 2. Topological isomers of a macrocycle.

by special methods in the electron microscope which can produce pictures which clearly reveal the topological nature of the bond holding the two specific macrocycles together. Since 1969 when two-component IPN's were first reported synthesized a large number of investigations[3-7] have appeared dealing with these materials which are of considerate interest since they can

produce blends of controlled miscibility of two crosslinked
networks of even relatively incompatible polymers. Indeed this
is the only way of achieving relatively miscible blends of
crosslinked, incompatible polymers. While graft bonds connecting
the different networks can be introduced inadvertently by side
reactions, one can also introduce grafting by design to produce
what are known as graft IPN's. IPN's can be formed from more
than two components. Recently mutually interpenetrating three
component IPN's have been reported, e.g. consisting of crosslinked polyurethane, epoxy, and polyacrylate networks[8].
Related to the full IPN's are the pseudo or semi IPN's in which
(in a two component case) only one of the networks is crosslinked
through which are interdispersed the linear chains of the other
component.

In the next section we describe the synthesis, phase stability
and what is known about the morphology of IPN's. In the final
section we comment on the observed physical properties in the
context of the chemical structure and the complex morphologies
exhibited by these materials.

2. Synthesis and phase morphology

IPN's can be basically formed: (1) In a sequential manner in
which a crosslinked network of type A polymer is swollen with the
monomer (1), crosslinking agent, initiator or catalyst for
polymer B which is formed and then crosslinked in situ to form a
"sequential IPN" or (2) In a simultaneous manner in which
monomer or prepolymer(s) of networks A and B together with their
specific crosslinking agents, initiators, catalysts or
accelerators are mixed (as latices, in solution or as neat
liquids) and cured to form what is called a SIN (simultaneous
interpenetrating polymer network).

Our focus here is on SIN's and we will give a number of
synthetic examples:

Example 1 Polyurethane acryclic copolymer IPN[9] (PU-PA IPN) A
polyurethane (PU) prepolymer (NCO/OH=2/1) is prepared by reacting
one equivalent of poly (1,4-oxybutylene) glycol, (M.W. = 1017)

with 2 equivalents of 4,4'-methylene bis(cyclohexyl isocyanate)
acrylate copolymer (AC) prepolymer is employed composed of eight
parts n-butyl methacrylate, one part of ethyl methacrylate and
one part of styrene. The PU prepolymer 1,4-butanediol and
trimethylol propane (crosslinking agent for the polyurethane),
0.24 (wt. of PU)% dibutyl tin dilaurate are mixed thoroughly with
the acrylate prepolymer syrup, 1 (wt. AC)% benzoyl peroxide and
glycol dimethacrylate (crosslinking agent for the AC
prepolymer). The mixture was cast between two glass plates
provided with rubber seals around the edges and polymerized at
110°C. for 24 hours. IPN's composed of 20, 40, 60, and 80% by
weight of PU could be thus made. The two different pseudo or
semi - IPN's can be similarly made by leaving out the
crosslinking agent for the appropriate network.
Example 2. Poly(2,6 dimethyl-1,4-phenylene oxide) (PPO) -
polystyrene (PSt) IPN[10].

PPO can be brominated with bromine or more cleanly with
N-bromosuccinimide to produce a PPO brominated at the methyl
group. This material can be mixed in solution with
ethylenediamine (with which it condenses splitting out hydrogen
bromide), styrene, divinylbenzene and a radical initiator to
produce the PPU-PST IPN of various compositions. Various
polyaddition, polycondensation, chain reaction polymerizations
have been employed to make thermosetting IPN's, as well as
thermoplastic IPN's[5] where the networks involved ionomeric
bonds and the crosslinking involved physical forces between
chains such as are involved in crystallization.

In example 1 above the IPN's showed heterophase behavior with
domain sizes ranging from 10-50 nm, apparent phase inversion at
60% PU by weight and two distinct Tg's. The Tg's as is often
observed lay between the Tg's of the pure crosslinked networks
and were shifted inward towards one another. The largest inward
shift (of over 10°K) occurred at the IPN with 80% PU. The
PPU-PSt IPN of example 2 above appears single phased down to the
resolution of transmission electron microscopy and exhibits a

single Tg varying monotonically between the Tg's of the pure
crosslinked networks.

The extent of miscibility, interpenetration, or phase
morphology depends not only on the thermodynamic condition, i.e.
the chemical nature of the reactants, solvents, additives,
temperature, pressure, etc. but also the chemical kinetic factors
such as dilution, catalysts, accelerators, temperature, pressure,
etc. which can affect the time development of the complex
reacting system producing the IPN. Thus, in example 1 the IPN is
produced from a relatively incompatible polymer network pair
while in example 2 the IPN's are formed from a network pair which
forms fully compatible, single phase, mixtures of the two linear
polymers over essentially all compositions. In example 1 by
changing the catalyst concentrations of the PU network or the
initiator concentration for AC network which changed the gel
times of the networks one could change the domain size as well as
shift the Tg's of the resulting IPN's[9,11,12]. To oversimplify
a description, if one could characterize the dominant time which
produces the most visible phase domain by uphill diffusion by the
half-life τ in an IPN composed of networks A and B whose
effective half-lives for formation and gelation are θ_A and
θ_B respectively a number of very different regions of
behavior can ensue: If the networks are very incompatible
thermodynamically then some phase miscibility is only possible if
$\theta_A \sim \theta_B = \theta$, the joint chemical half-life since
otherwise one network will squeeze out the molecules of the
second slower forming network. If $\theta > \tau$ one could not expect
much interpenetration except along the surfaces of the phase
domains which already will have formed. This would be consistent
with some of the models described in Sperling's monograph[5]. If
$\tau \gg \theta$ one could ask reasonable whether single microphase.
stable IPN's can be formed. Binder and Frisch[13] suggest, e.g.
that for chemically quenched IPN's, $\theta \sim 0$, whose networks are
sufficiently weakly crosslinked and are described by a Flory
statistical network theory even with a positive interaction

parameter thermodynamically stable phases would occur for all SIN IPN's (and pseudo IPN's) but no sequential IPN would be stable over all network compositions. Probably one of the most usual situations is one in which τ~Θ in which case the detailed prediction of the phase morphology becomes particularly difficult. In any case these considerations suggest that IPN morphologies can be highly varied ranging from truly molecularly homogeneous materials to interwoven sponges in which one network passes through the holes of the other to finally materials with one network dispersed in the other and the interpenetration restricted to a very thin interface region.

3. Physical properties

The primary question is how the specific chemical (or physical more or less) permanent entanglements of IPN's and their phase morphology affect the various physical properties including mechanical properties. If Θ>τ or if for some other reason interpenetration were restricted to a very thin interface region whose property values do not differ much from the pure component networks then one expects that the properties should not be very different from an "ideal two-phase medium" (See Figure 3 below)

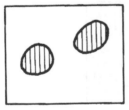 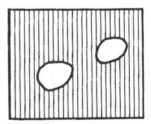

Figure 3. Schematic drawing of an ideal two-phase medium exhibiting phase inversion. Shaded region is phase A.
In such a medium one phase (say A) would be wholly dispersed in the other network phase and only above a certain B concentration would a phase inversion ensue. The weak interpenetration would be restricted to the interface. One expects that <u>all</u> properties

P of such an ideal two phase medium would be monotone functions
of composition, cf. Figure 4).

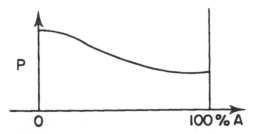

Figure 4. Schematic variation of properties P as a function of
composition for an IPN well approximated by an ideal two phase
medium.

The properties P of such IPN's would satisfy in good
approximation the predictions of mean field theories and would be
strictly bounded by properties of series and parallel
combinations of the two network phases. Indeed one sees many
examples of such behavior for mechanical and thermal properties
of many, IPN examples cited in Sperling's monograph[5]. But many
IPN's reflect their nature much more spectacularly when their
morphologies no longer can be well approximated by Figure 3. In
Figure 5 we show some examples of more complex behavior.

Figure 5. Schematic representation of more complex IPN
morphologies.

Departures from two-phase medium behavior can occur if
interpenetration results in a third phase (see left diagram in
Figure 5) particularly large departure occupy if some property of
that new phase is larger or smaller than those of the two pure
network polymers. These effects can of course be strongly
enhanced by complex morphological situations (shown on

the right of Figure 5) where no phase can be said to be properly
dispersed in any other. Now it might be possible to find some
property P which exhibits an intermediate maximum as a function
of composition. This is illustrated in Figure 6.

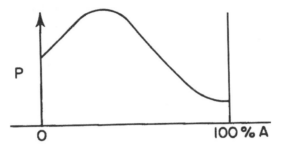

Figure 6. Schematic variation of some property of an IPN as a
function of composition with morphologies shown in Figure 5.

A number of SIN-IPN systems show this behavior. The PU-PA IPN
of example 1 of the previous section exhibits a dynamical complex
molecules which is shown in Figure 7 as a function of PA
concentration.[9]

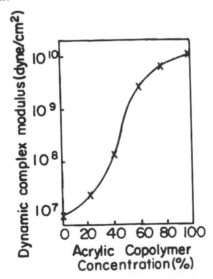

Figure 7. Dynamic complex modulus versus acrylic copolymer
concentration of the PU-PA IPN at 23°C. at a frequency of 110
Hz. (Crosses experimental points).

This variation of the modulus with composition is exactly what is expected of a property of an ideal two phase medium and like other properties such as density, etc. The solid line is a fit a Budiandsky equation[14] (a mean field theory) capable of representing the modulus response of a system exhibiting phase inversion to the experimental data. On the other hand the tensile strength to break (at room temperature, crosshead speed of 2 in/min.) and elongation at break showed maxima as a function of composition as shown in Table 1 below[9]:

Table 1 Physical properties PU-PA-IPN

Property	PU	100%	80	60	40	20	0
	PA	0%	20	40	60	80	100
Tensile strength (psi)		6100	7100	6525	5000	3715	2571
Elongation (% at break)		640	780	540	270	180	15
T_g(°K)		204	214	211	209	210	320.5
		–	306	306	308	312	–

The effect of catenation on the tensile strength may reflect the extra bonds due to the permanent entanglements which have to be broken and which appears to largest at 80% PU where one finds the largest inward shifts of the Tg's.

A similar maximum in tensile strength to break is observed at 25% PSt in the PPO-PSt IPN of example 2, cf. Figure 8 below[10].

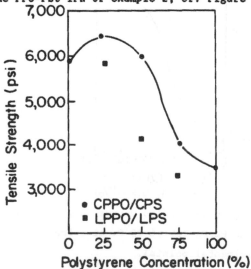

Figure 8. Tensile strength to break at room temperature versus
polystyrene concentration for the PPO-PSt IPN (filled circles) and
the miscible blend of the two linear polymers (filled squares).

Polyurethane-poly (epoxy) (PU-PE) IPN's exhibit a <u>maximum</u> in lap
shear versus composition and a <u>minimum</u> in the critical surface
tension versus composition at the same concentration that their
tensile strength to break exhibits a maximum.[15] This is shown in
Figures 9 and 10 where IPN I refers to a PU network using polyurethane prepolymer made from poly (1,4-oxybutylene) glycol while
IPN's II and III use a polyurethane prepolymer made from poly
(caprolactone) glycol of molecular weights 530 and 2000
respectively[15].

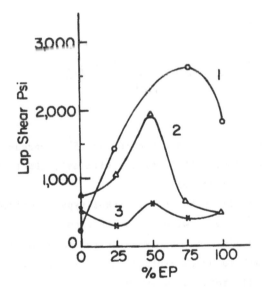

Figure 9. Lap shear (psi) versus epoxy network composition. IPN-I
(open circles), IPN-II (triangles), IPN-III (crosses).

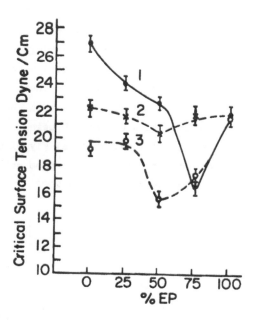

Figure 10. Critical surface tension at 25°C. versus epoxy
network composition. IPN-I (filled circles), IPN-II (unfilled
circles), IPN-III (crosses).

Permeability coefficients of this IPN-I type described above while
monotonically[15] varying with PU composition for toluene and water
vapor exhibit a minimum for oxygen at the same composition that the
tensile strength to break exhibits a maximum value.[16]

Tg's larger (in some cases by 50°K.) than either component
network could be observed for certain compositions of SIN
polydiacetylene-epoxy IPN;s as well as sequentially prepared IPN's
from two different polydiacetylenes.[17]   Recent work on polymeric
catenanes[18], in which large (92 repeating unit) cyclic
polysiloxanes are threaded by crosslinked poly (2.6 dimethyl 1,4-
phenylene oxide) clearly confirms that such topological isomeric
structures have unusual properties derived from their permanent
entanglement.[19]

In conclusion we see that under certain conditions IPN formation
results in novel materials some of whose properties are much more
complex than those expected of ideal two phase media.

ACKNOWLEDGEMENT

This work was supported by the National Science Foundation under
Grants DMR8305716A and DMR8515519.

REFERENCES

1. H.L. Frisch and E. Wasserman, J. Am. Chem. Soc. 83, 3789(1961).

2. H.L. Frisch and D. Klempner, Adv. Macromol. Chem. 2, 149 (1970).

3. O. Olabisi, L.M. Robeson and M.T. Shaw, "Polymer-Polymer
   Miscibility", Academic Press, New York, 1979.

4. Yu. Lipatov and L. Sergeeva "Vzaimopronikayushchie setki"
   (Interpenetrating Polymer Networks), Naukova Dumka, Kiew, 1979
   (in Russian).

5. L.H. Sperling "Interpenetrating Polymer Networks and Related
   Materials", Plenum Press, New York, 1981.

6. K.C. Frisch, D. Klempner, H.X Xiao, E. Cassidy and H.L. Frisch,
   Polymer Eng. Sci. 25, 758 (1985).

7. H.L. Frisch, British Polymer Journal. 17 (2), 149 (1985).

8. E.F. Cassidey, H. X. Xiao, K.C. Frisch and H.L. Frisch, J.
   Polym. Sci. (Chem. Ed.).

9. D. Klempner, H.K. Yoon, K.C. Frisch and H.L. Frisch in "Polymer
   Alloys II" (ed. by D. Klempner and K.C. Frisch), Plenum Press,
   New York, 1979, p. 185 see also H.L. Frisch, K.C. Frisch and D.
   Klempner, Pure & Appl. Chem. 53, 1557 (1981).

10. H.L. Frisch, D. Klempner, H.K. Yoon and K.C. Frisch,
    Macromolecules 13, 1016 (1980).

11. L.H. Sperling and R.R. Arnts, J. Appl. Polymer Sci. 15, 2731
    (1971).

12. R.E. Touhsaent, D.A. Thomas and L.H. Sperling, J. Polymer Sci.
    46C, 175 (1974).

13. K. Binder and H.L. Frisch, J. Chem. Phys. 81, 2126 (1984).

14. B. Budansky, J. Mech. Phys. Solids 13, 223 (1965).

15. H.L. Frisch and K.C. Frisch Progress in Organic Coatings $\underline{7}$, 105 (1979) and references cited therein.

16. S.A. Chen and H.L. Ju, J. Appl. Polym. Sci. $\underline{25}$, 1105 (1980).

17. P.V.S. Ika, H.L. Frisch, G.R. Walters, P.C. Painter and K.C. Frisch, J. Polym. Sci. (Chem. Ed.).

18. D. Callahan, H.L. Frisch and D. Klempner, Polym. Eng. and Sci. $\underline{15}$, 70 (1975).

19. T.J. Fyvie, H.L. Frisch, J.A. Semlyen, S.J. Clarson and J.E Mark, J. Polym. Sci. (Chem. Edit).

NITRILE RUBBER-MODIFIED POLY(ε-CAPROLACTAM) BY ACTIVATED ANIONIC POLYMERIZATION: SYNTHESIS, MOLECULAR CHARACTERIZATION AND MORPHOLOGY.

Giovanni Carlo Alfonso, Giovanni Dondero, Saverio Russo, Antonio Turturro
Centro Studi Chimico-Fisici di Macromolecole Sintetiche e Naturali, C.N.R., Corso Europa, 30, 16132 Genoa, Italy

and

Ezio Martuscelli
Istituto per la Tecnologia dei Polimeri e Reologia, C.N.R., Via Toiano, 2, 80123 Arco Felice (Naples), Italy

SUMMARY

The activated anionic polymerization of ε-caprolactam in presence of butadiene-acrylonitrile elastomers has been extensively studied. The classical, relatively slow activators lead to monomer conversion and polymerization rates that are, to some extent, reduced in comparison to polymer formation from the neat monomer. Moreover, the polymerization kinetics is dramatically affected by the degree of mixing of the reactants containing the catalytic species, the monomer and the rubber, as well as the contact time between the basic initiator and the rubber.
Very fast polymerization activators minimize the retardation effects of the rubber, thus suggesting possible developments of the polymerization process in terms of RIM technology. By suitably changing the experimental procedures it is possible to induce controlled crosslinking and degradation of the rubber, which represent essential prerequisites in order to obtain high monomer conversions as well as peculiar morphologies of the multiphase polymer system. Molecular characteristics and morphologies of the polymerization products have been correlated to the most relevant reaction parameters.

INTRODUCTION

In recent years several new materials, based on polymeric components, have reached a prominent role in many fields of great technological development. Among the various methodologies for the
achievement of such goals the most successful ones are usually based on intimate blending of two or more polymers, sometimes together
with additional properties-modifiers , in order to attain considerable enhancement of the specific properties of the parent components. Thus, the study of polymer blends has involved an increasing
number of research groups, devoted to this topic and related areas,
both at academic and industrial levels.

One of the most important polymer materials, suitable for extrusion and molding, is poly(ϵ-caprolactam) (PCL), which exhibits
several physical and mechanical properties worth to classify it as
an engineering plastics. Moreover, some of its intrinsic characteristics, such as toughness or fire resistance, can in principle be
enhanced or modified by the presence of suitable additives. However,
the usual compounding techniques, based on the melt mixing of the
polymer with a rubber or a flame-retardant agent, can cause severe
technological problems, mostly due to the relatively high processing
temperatures (ca. 270°C) and related to either or both the poor
thermal stability of the property-modifiers and the handling of dan-
gerous or toxic products.

We have recently found that PCL-based materials with improved
properties and performances can easily be obtained by activated a-
nionic polymerization of ϵ-caprolactam (CL) in presence of adequate
amounts of various commercial additives, provided this do not adver-
sely affect the polymerization reaction. By using this procedure
('in situ'mixing by polymerization) it is possible to bypass the afore -

mentioned processing problems; the only stringent requirement is
that the additive should not exhibit any adverse effect on thermo-
dynamics, kinetics and mechanisms of the polymerization. In this
respect, successful results have been recently attained[1-4] in the
framework of our research project on activated anionic polymerization of CL in presence of various additives[5-14]

Aim of the work.

Attempts to polymerize CL added with various fillers, reinforcing agents and property-modifiers have been successfully exploi-
ted, with a view to synthesizing relatively large polymer blocks,
containing the additive, dissolved or finely dispersed in the monomer medium, which remains homogeneously distributed in the polymer
matrix. Indeed, if the anionic polymerization reaction is not ad-
versely biased by the property-modifier, the very short polymerization time and the rapid increase of the medium viscosity prevent
any coarse aggregation of the additive and the consequent phase
separation in large domains.

Very recently, the activated anionic polymerization of CL has
been found suitable for the reaction injection molding (RIM) technology[15-18] and has prompted several new studies aimed to update the
classical picture of the polymerization kinetics, in the perspecti-
ve of industrial applications of the RIM process[14,19-22]. As compared to the well-known methodology based on the activated anionic
polymerization of CL[23](the so-called 'cast nylon' technology), the
RIM process introduces two new and severe requirements:

i) very fast polymerization kinetics, with 'equilibrium' conver-
    sion to be reached in less than 60 s.

ii) attainement of impact-resistant PCL, especially at low tempera-
    tures and in dry conditions, with other characteristics unaf-
    fected or improved.

The above requirements are consequences of specific economical
and technological demands, respectively, in order to produce new
materials with an optimum balance between cost and performances.
Thus, suitable property-modifiers have to be added to CL monomer
and new classes) of very active polymerization catalysts (initiator
and activator used in order to accelerate the reaction kinetics.

Indeed, it should be recognized that the anionic polymerization of CL, as a whole, is constituted of a complex set of main and
side reactions, which - under specific experimental conditions -
can produce a large variety of active species and irregular structures in the polymer chains[23]. Thus, the RIM process for materials
based on PCL requires a careful control of all parameters, that
can affect the pattern of the activated anionic polymerization
as well as the properties of the resultant materials. Our research
activity in this field has been focused on the following subjects:

i)   Role of activator and initiator concentrations in determining
     the optimum conditions for the polymerization of neat CL in
     an adiabatic reactor and in the mold[14,24].

ii) New catalysts for a very fast polymerization reaction[25].

iii) Impact-resistant PCL by dissolution of a suitable rubber in
     CL monomer, followed by rapid polymerization[1,2,24,25].

iv) Flame resistant PCL by anionic polymerization of CL in pre-
     sence of proper flame-retardant agents[3,4,26].

v)   Activated anionic polymerization of CL in presence of rein-
     forcing agents[24,25] (R-RIM).

The results which will be herein outlined refer to theme iii)
i.e. to our studies devoted to improve toughness of PCL, especially at low temperatures and in dry conditions. Unlike other research
groups, who have privileged multi-block copolymer formation between PCL sequences and the rubbery component in order to achieve

good impact resistance[17,18,21,22], we have not used any telechelic prepolymer, but simply dissolved an 'unreactive' elastomer in CL monomer. The meaning of the inverted commas will become clear in due course.

Among the various polar rubbers which can be more or less easily dissolved in molten CL at temperatures suitable for the activation of the polymerization reaction, i.e. from 120 to 160°C, in the present study we have paid special attention to high molecular mass butadiene-acrylonitrile rubbers.

EXPERIMENTAL

Materials.

Special grade CL monomer with water content lower than 100 ppm and commercial butadiene-acrylonitrile rubbers (Europrene type) were kindly supplied by Enichimica. The molecular characterization of the three rubber samples used in our experiments (E 3325, E 3345, and E 3945), performed in their research laboratories, is given in Table 1.

Table 1.

Molecular characteristics of Europrene rubbers

Type	AN, wt. %	$\bar{M}_v \cdot 10^{-5}$	$\bar{M}_w \cdot 10^{-5}$	$\bar{M}_n \cdot 10^{-5}$
E 3325	33	1.07	1.60	0.37
E 3345	33	1.44	1.62	0.51
E 3945	39	1.40	1.60	0.37

The above rubbers were extensively washed with hot water and carefully dried prior to use. By dissolution in acetone, a very low gel content was found in all samples.

Polymerization and samples preparation.

The 'slow' anionic polymerization of CL, containing various amounts of dissolved nitrile rubber (ranging from 5 to 12.5 wt.%), was carried out as follows: initially, 2/3 of the total amount of CL was introduced in a quasi-adiabatic reactor (total volume = 200 cm^3) together with the nitrile rubber. The temperature was raised to 150°C and the mixture kept under mechanical stirring in a dry nitrogen atmosphere for the time necessary to complete disso- lution of the rubber (1-3 hrs). The activator (usually, 0.6 mole % of N-acetylcaprolactam, referred to 100 moles of total CL) was then added up. Finally, a solution of initiator (usually, 0.6 mo- le % of sodium caprolactamate) in the remaining 1/3 of CL at T = 150°C was added under stirring and the polymerization reaction immediately started. Initiator and activator were prepared as pre- viously described[14]. The thermokinetic analysis of the polymeriza- tion runs was performed on the basis of the heat balance given in Ref. 7. By water extraction in Soxhlet for 48 hrs, the polymeriza- tion products were fractionated into high polymer (PCL plus rubber) and water-soluble fractions (containing residual CL monomer, higher cyclic oligomers, catalytic residues and low molecular mass side products). Subsequent Soxhlet extractions with acetone (48 hrs), in order to dissolve the original nitrile rubber, and hexafluoro- butanol (HFB) (24 hrs), in order to selectively extract PCL[27] from the acetone residue, enabled to evaluate their respective yields.

Simulation experiments.

The effects of temperature and basicity of the medium on ni- trile rubbers stability were evaluated by simulation runs at conditions corresponding to those used in actual polymerization experiments, followed by the analytical procedure outlined above. Namely, in run a) a solution of the rubber in CL (10 wt.%) was kept at 150°C for additional 30 min and then rapidly cooled at room temperature. The rubber was recovered and characterized. In run b) the rubber solution, after addition of 0.6 mole % of sodium caprolactamate, was kept at 150°C for 30 min. After cooling, an extraction and characterization procedure analogous to that of run a) was adopted.

Molecular characterization.

Original rubbers and acetone-extractables were characterized by infrared spectroscopy on KBr discs and viscometry in acetone at 20°C. HFB extractables and residues were characterized by infrared spectroscopy. No detailed analysis was carried out on water-extractables. Residual monomer content was evaluated by vacuum evaporation of finely ground material at 120°C. GPC analysis on PCL component was performed on N-trifluoroacetylated samples[28,29]. With 'fast' activators (bis-carbamoyl-lactams) a similar procedure was adopted for the synthesis and the mass balance of the reaction products, as well as their molecular characterization. The major difference consists in the inversion between activator and initiator; the latter was added first to the elastomer, in order to ensure some crosslinking of the rubber even in very fast polymerization conditions.

Morphological analysis.

Morphological characterization was performed by scanning electron microscopy (SEM) on cold-fractured, gold-coated, surfaces of the samples. A Cambridge Stereoscan 250MK2 apparatus was used. Observations

were carried out on as-polymerized materials, as well as specimens
etched with acetone and HFB.

RESULTS AND DISCUSSION

Three different types of nitrile rubbers, characterized as described in Table 1, were used in order to establish the effects of
molecular mass and chemical composition of the elastomer on polymerization kinetics as well as   properties of the resultant material.
The initial rubber concentration in CL was varied up to 12.5 wt.%.
Other operational variables which were taken in consideration include relative and absolute concentrations of initiator I, and activator A, and stirring time during polymerization, t_s.

The reaction parameters which have been considered the most relevant for our study were the overall time of polymerization, t_p,
the monomer conversion, λ, and the high polymer yield, P.

Slow activation.

For samples synthesized using N-acetyl CL as 'slow' activator only an
intimate mixing of the various components of the reacting mixture
provides optimum results. Thus, prolonged stirring after mixing of
the two solutions (one containing rubber, CL and activator, the
other CL and initiator, as described in the experimental) is an essential pre-requisite for the obtainement of the highest conversion
values in the shortest polymerization times.

As an example, the effect of stirring time on three different
polymerizing systems is compared in Table 2. For neat CL, a sharp
reduction of t_s causes a considerable increase of t_p (+ 30%), whereas P remains unaffected; for the mixture containing 5 wt.% of
E 3345 rubber, both t_p and P are strongly biased (+ 44% and -1%,
respectively); this behavior is fully confirmed by the system

containing 5 wt.% of E 3945, where t_p increase is more then 13% and
P decrease is about 4%. Many other tests in various experimental
conditions fully confirm these differences and the relevance of stirring time on polymerization kinetics.

From the above results we can deduce that two different factors
for the systems with and without rubber play an important role in
the polymerization reaction, when t_s is strongly reduced. On one
hand, data from neat CL polymerization show that P, even when t_s is
lowered to only 10 s., coincides with its limiting conversion value
related to that concentration of active species.

TABLE 2.

Effect of stirring time on polymerization kinetics and polymer yield
$([I]/[A] = 0.6/0.6 \text{ mole}\%)$

Sample	rubber, type and content, wt. %		t_s, s.	t_p, s.	P, %
L38	none		150	300	95.3
L39	none		10	390	95.5
L25	E 3345	5	240	480	94.2
L27	"	"	30	690	93.3
L34	E 3945	5	300	450	95.0
L15	"	"	10	510	91.0

As described in detail in Ref. 14, only when equimolar concentrations of initiator and activator are used, maximum values of monomer
conversion and high polymer yield are reached; thus, it can be inferred that the overall equimolar stoichiometry between initiator and
activator is well preserved throughout the system. The retardation
effect for the sample L39 is simply due to the time necessary for

the active species to diffuse and balance their concentrations also
locally.

On the other hand, when a high molecular mass rubber is dissolved in molten CL, the intimate mixing of this very viscous solution
with the low-viscosity one, containing the initiator dissolved in
the remaining CL, is hardly attained. Only adequate stirring, prolonged during the initial period of polymerization, can assure mixing between the active species, thus providing short polymerization
time and optimum conversion. The remarkable viscosity effects cause
severe limitations to the mutual diffusion of initiator and activator molecules, as well as   growing chains and elastomer, when stirring is too short. As a consequence, not only t_p, λ and P are strongly affected, but also substancial rubbery phase separation during
polymerization occurs. Direct evidence is shown by the micrograph of
Fig. 1. When t_s is less than 60 s., the morphology of the system is
characterized by large domains of both neat and rubber-coated PCL.

For the mixtures containing various amounts of nitrile rubber
any variation of the [I]/[A] ratio below unity causes an increase
of the polymerization time and a sharp decrease of the high polymer
yield, as shown in Table 3. However, at variance with previous fin-
dings related to the neat system, an [I]/[A] ratio higher than unity
shows some improvements of both t_p and P%. The above data suggest
an active role of the nitrile rubber in this connection, as will be
discussed in more detail further on. These effects overcome the variations of the stirring times for the various samples.

For samples synthesized under prolonged stirring (150 s.) after
the start of polymerization, t_p, λ and P as functions of the initial rubber content are given in Fig. 2. Both initiator and activator concentrations, [I] and [A], are 0.6 mole %, referred to 100 mo-

les of CL.

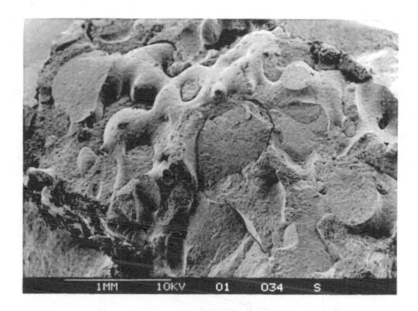

Figure 1.- SEM cold fractured surface of PCL/E 3325 95/5 blend; N-acetyl-caprolactam as slow activator and stirring time less than 30 s.

TABLE 3

Effect of [I]/[A] ratios on polymerization kinetics and polymer yield (t_s between 60 and 300 s.)

Sample	rubber, type and content, wt.%		[I]/[A]	t_p, s.	P, %
L38	none	none	0.6/0.6	300	95.3
L36	E 3325	5	0.6/0.6	390	93.3
L 7	E 3325	5	0.2/0.6	900	69.7
L25	E 3345	5	0.6/0.6	480	94.2
L17	E 3345	5	1 /0.6	450	95.9
L33	E 3345	10	0.6/0.6	960	91.8
L12	E 3345	10	1 /0.6	910	93.0

As compared to the values pertaining to the polymerization of
neat CL (t$_p$ 300 s., λ 98%, P 95.3%), it can be seen that all types
of rubber cause retardation to the polymerization reaction as well
as a sharp decrease of both monomer conversion and high polymer yield. This behavior cannot be attributed to the rather small changes
in activator, initiator and monomer concentrations, i.e. to the dilution effects due to the solubilized rubber. Indeed, on the basis
of our previous results[14], the overall effects of dilution on t$_p$,
λ and P can be represented by the dotted lines given in each plot.

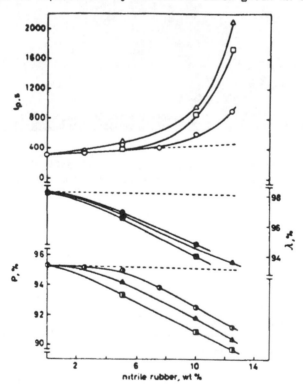

Figure 2.- Effect of rubber type and content on CL polymerization.
△,▲, ◭ : E 3345; ☐,◼, ◨ : E 3325; ○,●,◓ : E 3945

The retardation and the decrease of λ and P are well outside these
lines and directly related to the rubber type and concentration.

These effects are less pronounced for the rubber with the highest
AN content (E 3945), which is also the most readily soluble in molten CL.

Both λ and P values are the lowest for E 3325-based runs, whereas t_p data are highest for E 3345-based ones. This inversion cannot be entirely attributed to differences in molecular mass distribution and averages of the original rubbers, taking also in account
that they undergo modifications during polymerization in terms of
crosslinking, as will be shown later on.

As shown in Fig. 3 the morphology of a typical system is constituted of a continuos PCL matrix containing large domains of va-
riable size.

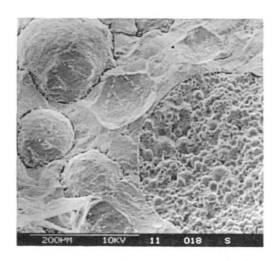

Figure 3.- SEM cold fractured surface of PCL/E 3325 90/10 blend;
N-acetyl-caprolactam as slow activator and stirring time
of about 250 s.

In turn these domains are made of PCL spherical sub-domains coated with a soft, rubbery phase. At the boundary between PCL matrix and the macro-domains, a large number of interconnecting bridges is evident.

This peculiar morphology is present for all systems studied, provided that prolonged stirring is ensured. Depending on rubber type and content, as well as   the polymerization kinetics, the macro-domains vary their size distribution,as listed in Table 4. Indeed, when the rubber is richer in AN, the morphology is characterized by smaller domains, at other parameters constant. By increasing its amount from 5 to 10 wt.%, the average dimensions increase from 34 to 73 μm. By more prolonged stirring, average dimensions decrease at about 40 μm even if the rubber content is further increased up to 12.5%.

The mass balance of the polymerization products has been evaluated by selective solvent extractions: water treatment in Soxhlet has enabled to determine high polymer (PCL) yield and overall CL oligomer concentration (up to the cyclic nonamer); the residual

TABLE 4

Effect of polymerization conditions on the bicomponent macro-domain size ([I] /[A] = 0.6/0.6 mole %)

Sample	rubber, type and content, wt.%		t_s, s.	t_p, s.	$\bar{\phi}$, μm
L25	E 3345	5	240	480	54
L34	E 3945	"	300	450	34
L35	"	10	300	600	73
AM26	"	12.5	520	900	40

product has been treated with acetone in Soxhlet in order to evaluate the removable portion of nitrile rubber. Only a fraction of its
initial content (from 25 to 78%, depending on the polymerization
conditions) has been extracted. The residue has undergone a further
extraction with HFB, a good solvent of both PCL and nitrile rubbers.

Also after HFB extraction, some residue remains in the thimble.
IR characterization of all extractables and residues reveals that:

i)   with acetone, the nitrile rubber is removed only in part;

ii) with HFB, pure PCL without any traces of rubber is extracted;

iii) the residue is entirely made of nitrile rubber.

Mass balances of extractables and residue always match the o-
riginal values before extraction.

Further characterization of the residue by calorimetric and
swelling techniques reveals that a moderately crosslinked nitrile
rubber is formed in the experimental conditions used in our synthesis. Indeed, a slightly higher T_g value and a high degree of swelling have been found. The molecular characterization data are fully confirmed by morphological observations. Fig. 4, related to the
acetone-etched surface of L36 sample (5 wt.% of E 3325), reveals
that, after removal of the soft, rubbery cover by acetone, the PCL
spherulites do not fall apart, but are kept together by the unso-
luble fraction of rubber.

PCL particle dimensions are very regular, with an average diameter of about 4 μm. On the other hand, Fig. 5, referring to the
HFB-etched surface of the same sample, clearly shows the continuous
network phase made of crosslinked nitrile rubber.

In order to ascertain whether the crosslinking reaction on the
nitrile rubbers is due to purely thermal effects or to the influen-

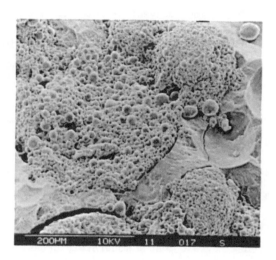

Figure 4.- Micrograph of PCL/E 3325 95/5 blend etched with acetone, solvent of the rubbery component.

ce of the reaction medium, simulation runs as previously described were performed. A comprehensive set of results is given in Table 5.

From these data it clearly appears that the major factor inducing the crosslinking reaction is the basicity of the medium. Together with crosslinking , also some degradation of the soluble fraction seems to occur. Indeed, a sharp decrease of the limiting viscosity number of the soluble fraction has been found for all rubbers

which underwent basic medium simulation experiments. In actual polymerization, the extent of crosslinking and degradation is reduced: for instance, 10% of E 3325 in the polymerizing system (sample

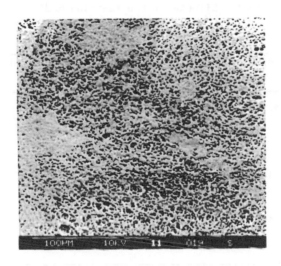

Figure 5.- Micrograph of PCL/E 3325 95/5 blend etched with hexafluo-robutanol, solvent of PCL

L32 in Table 5) gives 36% of crosslinking and 66 as limiting visco-sity number of the acetone-soluble fraction. It appears, therefore, that some competition between the polymerization of CL and the cros-slinking (and degradation) of the rubber takes place.

The extent of the crosslinking reaction during polymerization seems to be a complex function of the experimental conditions, as shown by the data given in Table 6. At the same rubber content, AN-richer elastomers have a higher crosslinked fraction (77.5% of L34 as compared to 40.7 of L33). For all types of rubbers the crosslin-ked fraction decreases as the rubber content increases, thus implyi-ng an active role of the initiator (and derived basic species) on the crosslinking reaction, in competition with the PCL growth me-chanism. Therefore, mutual effects of crosslinking extent, monomer

TABLE 5

Simulation experiments. Effects of the reaction medium on nitrile rubber stability (T = 150°C).

Sample	rubber, type and content, wt.%		run[a]	simulation conditions	%crosslink. rubber	$[\eta]^{b}$,cm^3/g
SL9A	E 3325	10	a	thermal	2.3	81
SL9	"	"	b	basic med.	48.0	40
L32	"	"	-	actual polym.	36.4	66
SL11A	E 3345	"	a	thermal	2.6	107
SL11	"	"	b	basic med.	60.0	33
SL16A	E 3945	"	a	thermal	3.5	92
SL16	"	"	b	basic med.	82.0	47

[a] see Experimental

[b] limiting viscosity number of the acetone soluble fraction. $[\eta]$ values for the original rubbers are 83, 100, and 99, cm^3/g for E 3325, E 3345 and E 3945, respectively.

TABLE 6

Effect of the experimental conditions on the extent of rubber crosslinking ($[I]/[A]$=0.6/0.6 mole %; t_s between 240 and 390 s.)

Sample	rubber type and content, wt.%		crosslinked rubber, wt.%	crosslinked fraction,%
L36	E 3325	5	3.28	65.5
L32	"	10	3.64	36.4
L25	E 3345	5	2.94	58.7
L33	"	10	4.07	40.7
L34	E 3945	5	3.88	77.5
L35	"	10	5.03	50.3
L44	"	12.5	4.58	36.6

conversion to polymer and overall polymerization time can be envisaged envisaged by comparing data from Table 6 and Fig. 2, even if more definite correlations are not possible with the set of data pre-
sently available.

The postulated mechanism of the crosslinking reaction for butadiene-acrylonitrile elastomers in basic media is based on the formation of carbanionic species of the type:

$$- CH_2 - \overset{\ominus}{\underset{\underset{CN}{|}}{C}} - \qquad\qquad (1)$$

presumably by direct proton exchange between the acrylonitrile mers
and initiator (sodium caprolactamate)or polyamide chain N-anions. This
exchange reaction is highly favoured when adjacent acrylonitrile
units are present, i.e. for AN-richer elastomers, where the AN-AN
diads concentration is higher.

The reaction of the carbanion (1) with the butadiene double
bond gives rise to inter- and intra-chain connections and, ultima-
tely, to crosslinking.

Degradation is originated by chain scission, as follows:

$$-CH_2-\overset{\ominus}{\underset{\underset{CN}{|}}{C}}-CH_2-\underset{\underset{CN}{|}}{CH}- \longrightarrow -CH_2-\overset{\ominus}{\underset{\underset{CN}{|}}{CH}} + \quad = CH_2 = \underset{\underset{CN}{|}}{C} - \qquad (2)$$

Similar reaction schemes have been recently proposed for the
alkaline degradation of styrene-acrylonitrile copolymers[30].

In actual polymerization runs, when initiator is purposely added first to the solution of rubber and caprolactam and contact time before introducing the activator is long enough to ensure complete crosslinking (about 1800 s.) the polymerization time is higher than that obtained with the usual procedure. Comparative re-

sults are given in Table 7 where sample SM4, characterized by complete crosslinking of the nitrile rubber, shows higher t_p and slightly lower P as compared to sample L34, with 77.5% of crosslinked rubber. More striking differences appear when mixing time (by stirring) is reduced: sample L13 (fully crosslinked) has very high t_p and rather low P, as compared to sample L15 (37.5% crosslinking). It can be, therefore, inferred that the consumption of active species to produce a fully crosslinked rubber causes a greater retardation of the polymerization kinetics, with somewhat minor effects on the high polymer yield.

The role of rubber type and content on PCL molecular mass and its distribution is given in Table 8.

TABLE 7

Effect of degree of crosslinking on polymerization kinetics and high polymer yield (type and content of the rubber: E 3945, 5 wt.%; [I]/[A]: 0.6/0.6 mole %)

Sample	t_s, s.	t_p, s.	P,%	% crosslinking
L34	300	450	95.00	77.5
SM4	240	870	94.33	100.0
L15	10	510	91.00	37.5
L13	10	1500	88.90	100.0

It can be noticed that $\overline{M}_n$ and $\overline{M}_w$ values show only slight variations, as compared to the corresponding data for neat PCL. There is almost no effect of E 3325 (up to 12.5 wt/%) and a rather moderate increase of average molecular masses in presence of E 3945, without any straightforward correlation with rubber content. In this

respect, it can be assessed that the elastomer plays an 'inert' role on the fate of polyamide chain (initiation, growth and termination).

TABLE 8

Effect of nitrile rubbers on PCL molecular mass and its distribution

Sample	rubber type and content, wt%		$\bar{M}_n$	$\bar{M}_w$	Q
A86	none		25100	51360	2.05
AM25	E3325	12.5	26390	53230	2.02
AM2	E3945	5	29280	59900	2.05
AM26	"	12.5	27820	61460	2.21

Fast activation.

Fast polymerization activators of the bis-carbamoyl lactam type, recently synthesized in our laboratory, have been tested in order to evaluate applicability of the RIM technology to the system
PCL-nitrile rubbers. Significant results of their effects on t_p, λ
and P are represented in Figure 6. It can be seen that the considerable increase of the polymerization rate for the neat system,
with a fivefold decrease of t_p, is well preserved also in presence
of large amounts of nitrile elastomers: t_p is ~200 s. for systems
with 10 wt% of rubber. For the sake of comparison, data related to
the same type of rubber and slow activation are also given. It can
be noticed that the two t_p curves are almost parallel: fast activators as compared to slow ones lead to a t_p shortage of 250-300 s.

In terms of λ and P the results clearly show a very favourable
trend, with a slight decrease of P well below the corresponding da-

ta related to slow activators. High P and λ values, together with
very short t_p times, ensure the suitability of the system, based on
nitrile rubbers dissolved in CL, to be utilized for RIM processes.

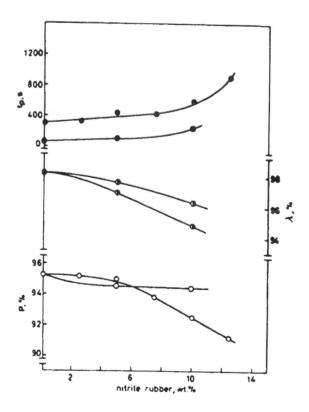

Figure 6. - Comparison between slow and fast activation. Effects on
t_p, λ and P.

O, ●, ◑ : slow activation;

O, ●, ◑ : fast activation.

Of course, stirring effects, very important for systems containing slow activators, cannot play any role for these 'fast' systems,
which induce rapid branching and crosslinking of the polyamide chains; thus, it is suggested that demixing phenomena can be success-

fully overcome with a polymerization process fast enough to cause a very rapid increase of conversion and, conversely, medium viscosity.

Preliminary morphological observations, on AM3 sample, containing 5 wt % of E3945 (Figure 7) reveal the same phase organization already depicted in Figure 3. However, peculiar differences between the two morphologies are due to the simultaneous presence of cylindrical and spherical macro-domains in the fast polymerized sample. A more detailed evaluation of the morphological characteristics of these 'fast' systems is under study and will be presented elsewhere.

In order to ensure the formation of crosslinked rubber structures, which are necessary to develop the unique morphologies and properties of the resultant material, an inversion in the polymerization procedure, already mentioned in the experimental, must be adopted for the 'fast activation' systems. The initiator is added

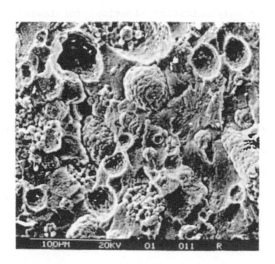

Figure 7. - SEM cold fractured surface of PCL/E3945 95/5 blend; fast activator of the bis-carbamoyl lactam type.

first to the CL solution of the elastomer and the contact is prolonged until adequate crosslinking of the rubber is achieved. Then,
the fast activator is introduced in the reactor and the polymerization starts.

CONCLUDING REMARKS

From the above data it is evident that the activated anionic
polymerization of CL in presence of high molecular mass nitrile rubber is a considerably complex reaction, characterized by strong mutual interactions among the various parameters. The polymerization
initiator acts also as a strong base on the nitrile rubber, causing
its crosslinking and partial degradation. The rubber network remains
suspended in the medium, preferentially swollen by the uncrosslinked rubber, as clearly shown by the morphological observations. Retardation of polymerization kinetics is due in part to the rubber
itself and, in part, to the crosslinked fraction. Conversely, any
variation of the polymerization kinetics, for instance by changing
[I], [A] or their relative ratio, remarkably affects not only the
extent of crosslinking but also the rubber distribution in the PCL
matrix at the end of polymerization. Thus, morphology and properties of the resultant materials are kinetically controlled by the
course of polymerization and strongly depend on the experimental
conditions chosen for the synthesis. The use of fast activators
(bis-carbamoyl lactams) allow to reach high conversions in rather
short reaction times; therefore, the RIM technology can be considered suitable for our system.

ACKNOWLEDGEMENTS

Is is a pleasure to acknolewgde the multi-year financial support from Progetto Finalizzato Chimica Fine e Secondaria-Sottopro-

getto Materiali Polimeri - Tematica Bb - C.N.R.

The authors are grateful to Dr. G.Costa, Prof. E.Pedemonte and Prof. G.Pezzin for helpful discussions and suggestions.

Thanks are due to Dr. A. Lagostina for help in the experimental part and Mrs. L. Rosini for the final editing of the manuscript.

REFERENCES

1. S.Russo, G.C.Alfonso, E.Pedemonte, A.Turturro and E.Martuscelli,
   Eur.Pat.Appl. (to C.N.R.) 84105456.2, (1984).

2. S.Russo, G.C.Alfonso, A.Turturro and E.Pedemonte, Eur.Pat.Appl.
   (to Enichimica) 84201713.9, (1984).

3. G.C.Alfonso, S.Russo, E.Pedemonte, A.Turturro and C.Puglisi,
   Eur.Pat.Appl. (to C.N.R.) 84106053.6, (1984).

4. G.C.Alfonso, G.Costa, S.Russo, A.Ballistreri, G.Montaudo and C.
   Puglisi, It.Pat. (to C.N.R.) 23115A, (1984).

5. G.Bontà, A.Ciferri and S.Russo, in: Ring Opening Polymerization,
   ACS Symposium Series, no. 59, p. 216, T.Saegusa and L. Goethals
   (Eds), A.C.S., Washington, D.C. (1977).

6. G.Costa, E.Pedemonte, S.Russo and E.Savà, Polymer 20, 713 (1979).

7. G.C.Alfonso, G.Bontà, S.Russo and A.Traverso, Makromol.Chem.
   182, 929 (1981).

8. G.C.Alfonso, E.Pedemonte, S.Russo and A.Turturro, Makromol.Chem.
   182, 3519 (1981).

9. E.Biagini, B.Pedemonte, E.Pedemonte, S.Russo and A.Turturro,
   Makromol.Chem. 183, 2131 (1982).

10. G.C.Alfonso, G.Cirillo, S.Russo and A.Turturro, Eur.Polym.J.
    19, 949 (1983).

11. G.C.Alfonso, G.Costa, E.Pedemonte, S.Russo and A.Turturro, Proc.
    5th AIM Meeting, AIM, Milan (1981).

12. S.Russo, Proc.IUPAC 28th Macromolecular Symp., Amherst (1982).

13. E.Biagini, S.Razore, S.Russo and A.Turturro, ACS Polymer Pre-
    prints 25, 208 (1984).

14. G.C.Alfonso, C.Chiappori, S.Razore and S.Russo,in: Reaction
    Injection Molding, ACS Symposium Series, no. 270, p. 163, J.E.
    Kresta (Ed.), A.C.S., Washington D.C., (1985).

15. R.S. Kubiak, Plast.Enging. 36, 55 (1980).

16. R.S.Kubiak and R.C. Harper, Reinf.Plast./Composites Inst., 35th
    Annual Conf., Sec. 22-C, 1 (1980).

17. R.M. Hedrick and D.Gabbert, A.I.Ch.E. Symp., Detroit (1981).

18. D.Gabbert and R.M. Hedrick, A.I.Ch.E. Symp., Detroit (1981).

19. R.E.Sibal, R.E. Camargo and C.W. Macosko, Int. J. Polym.Techn.
    Eng. Polym. Process Enq., 1, 147 (1983-84).

20. R.E.Camargo, V.M.Gonzales, C.W.Macosko and M.Tirrell, Rubber
    Chem.Techn., 56, 774 (1983).

21. R.M.Hedrick, J.D.Gabbert and M.H.Wohl,in: Reaction Injection
    Molding, ACS Symposium Series, no. 270, p. 135, J.E.Kresta (Ed.),
    A.C.S., Washington, D.C. (1985).

22. J.L.M.van der Loos and A.A. van Geenen,in: Reaction Injection
    Molding, ACS Symposium Series, no. 270, p. 181, J.E.Kresta (Ed.),
    A.C.S., Washington, D.C. (1985).

23. J.Sebenda, in: Comprehensive Chemical Kinetics, p.379, C.H.Bam-
    ford and C.F.Tipper (Eds.), Elsevier (1976).

24. Research under Contract CNR-ANIC S.p.A. reg. 22/12/1982.

25. Research under Contract CNR-ENICHEM Polimeri

26. G.C.Alfonso, G.Costa, M.Pasolini, S.Russo, A.Ballistreri, G.Mon-
    taudo and C.Puglisi, J.Appl.Polym.Sci. 31, 1373 (1986).

27. G.Costa, P.Bonardelli, G.Moggi and S.Russo, J.Macromol.Sci.-Chem.
    A18, 299 (1982).

28. E.Biagini, E.Gattiglia, E.Pedemonte and S.Russo, Makromol.Chem. 184, 1213 (1983).

29. E.Biagini, G.Costa, A.Imperato, E.Gattiglia and S.Russo, Polymer in press.

30. F.Severini and M.Pegoraro, Angew.Makromol.Chem. 117, 145 (1983).

REACTIVE EXTRUSION

G. Menges, T. Bartilla, P. Heidemeyer
Institut für Kunststoffverarbeitung, Aachen, West Germany

Introduction

Since about 20 years the extruder is used for the performance of
chemical reactions in the manufacturing and processing of polymers. Machines of that kind are used in both, the polimerization
as well as the modification of polymers - for example the graft
copolymerization. In addition to chemical changes, the selective
modification of the property spectrum of plastics is achieved
through mixing of polymers. In this context - the production of
blends - the extruder also opens up a wide range of possible
applications.

At the end of the sixties, Illing must have been the first to use
the extruder for the polymerization of caprolactam into polyamide
6 /1,2/.

It is primarily the manufacturers of raw material - rather than
the processors of plastics - who are concerned with extruders as
reactors since years and use them for this purpose with great
suceess. Very recently a few publications have been made which
deal with this subject /17, 18, 19/. It is said in these papers
that altogether there are only three puplications in this field
while a total of 200 patents have been issued to the manufactures
of raw material.

Among other plastics, the super-tough Nylons made by Du Pont, for
example, are manufactured in a two stage process in an extruder.

129

In the first step EPDM is modified with maleic anhydride. The
reactive EPDM is then grafted on the terminal amine of the
polycaprolactam chains.

Reichhold Chemical produces a polyolefin called Polybond which is
modified with acrylic acid; it features extreme adhesive characteristics. It is produced on a single screw extruder. Subsequently
to the melt zone of the extruder, liquid acrylic-acid monomer is
added and intermixed with the polyolefin melt.

The reactive polystyrene (RPS) put on the market by DOW-Chemical
brought about a dramatic change in the situation. This material
falls into a new category of plastics. The processor can combine
it with other, differing plastics in an extruder by means of
reactive processing. This option opens up entirely new perspecti-
ves to the processors as a consequence; but it also brings about
the necessity to grapple with a completely new technolgy.

The ability of RPS to react chemically results form the oxazoline
groups coupled up with the styrene polymer backbone. It is also
said that combining polystyrene with the - so far incompatible -
polyolefins will become possible. This will open up the probably
most interesting application of the new material in the field of
coextrusion. The chemical connection of single sheet layers with
each other improves the properties of the extrudate - in comparison to physical bonds with primers - even if the processing
conditions are extreme.

Examples of several other applications using co-rotating, self
wiping, twin screw extruders for polymerization are shown in
Table 1.

All in all, these developments bring about that the manufacturers
of machinery are required accordingly to extend their offered
range of extruders to be used as reactors for the application of
reactive processing.

End Product	Starring Material	Type of reaction
Polyethylene terephtalate	Precondensate-bis(hydroxyethyl) terephtalate	Condensation
Polybutylene terephtalate	Precondensate-bis(hydroxybutyl) terephtalate	Condensation
Polyamide 6,6	Precondensate	Condensation
Polyacrylate	Iso and terephthalic acid, bisphenols	Condensation
Polyurethane	Diisocyanate, polyols, aromatic amines	Addition
Poylamide 6	Caprolactam	Addition
Polyoxymethylene	Formaldehyde, trioxane	Ionic
PMMA	Methyl methacrylate	Free radical
Copolymer	Hydroxypropyl methacrylate, 2-ethylhexyl acrylate	Free radical
Copolymer	Ethyl acrylate-acrylic acid	Free radical
Graft copolymer	Poylstyrene-maleic anhydride	Free radical
Water soluble copolyamide	2 unsaturated amides	Free radical and devolatilization

Table 1: Polymerizations on co-rotating, self-wiping,
        twin-screw extruders /18/

The IKV at the University of Aachen, as a research institute with
more than 375 supporting members out of the industry has paid due
reguard to this development. An optimization of the extrusion
process using the polymerization of caprolactam as an example and
the development of new variations of processing the polyamide that
has been polymerized in the extruder ist carried out. However the
most important target of this work is to work out rules and stra-
tegies for the lay-out of extruders for these cases of application.

The polymerization of caprolactam has been chosen as a first
example because this reaction is well known and industrially
applied in plastic processing since about 30 years. In addition to
the classic monomer casting techniques it has lately been used in
the manufacture of polyamide mouldings by means of reaction
injection moulding aswell. The latter is subject of intensive
research investigations at the IKV. A further aspect is that the
high-molecular polycaprolactam moulding compounds and products
polymerized in the extruder may very well become of importance.

The reactive extruder

So far a great variety of extruders have been used as reactors
for the polymerization and the polymer modification. Obviously
the screw kneaders made by Buss, Welding Engineers, for example,
are applied here which are used for compounding, but above all the
co-rotating twin screw extruders made by Berstorff and Werner and
Pfleiderer are used. For our investigations aswell a small twin
screw extruder of the ZSK type made by Werner and Pfleiderer
was used.

The segmented design of barrel and screw allows for adaption of
this machine to different tasks. This way it is possible to
configure screws consisting of, for example, several conveying
elements differing in number of flight, conveying direction and
pitch of the screw and kneading elements with different geometry
and conveying effect (Fig. 1) /7,8/.

The ZSK extruder is a system which is open in the longitudinal
direction. As opposed to the counter rotating twin screw extruder
the conveyance of material results from drag forces here. A number
of different conveying mechanisms can be distinguished according
to the consistency of the moulding compound. Freeflowing solid
particles and ultra-high viscous melts are pushed through the
barrel in form of an 8-shaped threaded nut without deformation or
internal flow.

Figure 1: Screw elements and barrel sections

Viscous fluids such as polymere melts are conveyed in drag flow
in a spatial helix shaped in form of an eight. Particularly the
melt in the range of the gusset is intensively mixed and
rearranged here. Each of the two intermeshing screws is scraped
out along a spherical curve by the flight of the other screw. Thus
a good self-cleaning effect is obtained and a narrow residence
time distribution results which almost matches that of the plug
flow in tubular reactors.

Effects such as plate-out are practically eleminated with this.
The residence time and its distribution to the single zones can be
influenced by using conveying and kneading elements which act in a
conveying, neutral or restricting way, according to their
geometry.

Similarly, the axial mixing effect of single kneader geometries
will also influence the width of the residence time distribution.

In addition to the good mixing characteristics a further advantage
of extruders of this type lies in the favourable ratio of surface
to volume of 0,5 m²/l; it allows for intensive heat exchange.
The surface to volume ratio exceeds that of a tank reactor by
more than two orders. The heat exchange coefficients aswell
are relatively high; they amount to some 50 to 150 kJ/m².

Besides these positive features of extruders of that kind,
however, special attention should be given to the influence of
shearing. On the one hand it shurely has a favourable effect
on the mixing homogeneity - hence on the uniformity of the
reaction - but on the other hand, the dissipative heating of the
melt entails enhanced reaction speeds in the extruder.
This effect is of particular significance for high viscosities
hence towards the end of a polymerisation reaction.

With many thermally and mechanically sensitive materials, shearing,
the dissipative heating resulting there of and residence of the
melt at high temperatures lead to degradation of the molecular
chains; this effect is generally not wanted despite the fact
that it brings about increased homogeneitis in the molcular weight.

Precise feeding of the accurate proportion of system components is
required for the reproducibility of the reaction to be performed
in the extruder. Therefore the metering systems were used according
to the set-up depicted in Fig. 2. They gravimetrically proportion
the system components (activator, catalyst and caprolactam) that
are free-flowing solids at room temperature.

Caprolactam and the activated anionic polymerization

Investigations into the polymerization of polyamide 6 in the
extruder date back to the 1960s. It was Illing /1,2/ who first
published proposals regarding the extruder and screw configuration, although these were not implemented on an industrial
scale.

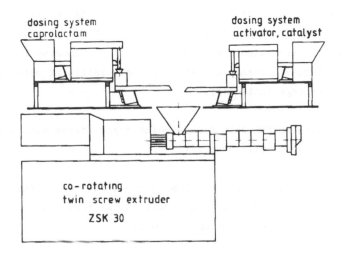

Figure 2: Assembly of the extrusion line to polymerize polyamide
6 in principle

Even at that time, the idea was to employ activated, anionically
reacting polyamide systems. These are currently used in monomer
casting /3/ and, in modified form, in the nylon RIM process as
well /4,5/.

It is basically possible to use standard commercial monomer
casting systems for polyamide 6 polymerization in the extruder,
providing that their reaction rate can be adapted to the residence
time of the extruder. The polymerization of lactams in the
presence of alkaline catalysts produces highmolecular polyamides
in just a few minutes, by contrast to "classical" polymerization
with reaction times of up to 10 hours /6,14,15/.

A typical representative of the lactams that react in this
so-called rapid polymerization is -caprolactam (Fig. 3, (1))

The catalysts used are substances which are capable of producing
lactam anions, e.g. alkali carbonates, alkali hydrides or
metallo-organic compounds /3,6/. After an induction period,
the polymerization takes place at a high rate. According to
/3,6/, the reaction commences with the ionic initiator attaching
itself ot the caprolactam (2).

A new, reactive sodium salt forms, which again reacts with any
caprolctam that is still free (3).

This reaction causes a reconversion of the initiator (sodium
lactam) and, at the same time, an acryl lactam forms, which
reacts very rapidly with sodium lactam (4).

The Na-salt that forms develops into a new initiator with a
caprolactam molecule and this, in turn, reacts with a further
acryl lactam.

It is possible for the reaction between the acryl lactam
(which has formed from the catalyst and the lactam at temperatures of above 190 °C) and the Na-lactam, and also for the
follow-up reactions to take place at low temperatures. Hence
acryl lactams and also acrylising compounds used as activators
will allow polymerization at low temperatures of approximately
120 °C /3,6/.

The investigations were conducted with a standard commercial
monomer casting system made by BASF. Tuning of the formulation
and limiting of the process parameters according to the polymerization in the extruder was carried out in the course of
the preliminary tests /9/ in a discontinously operating laboratory
casting-plant furnished with a glass reactor.

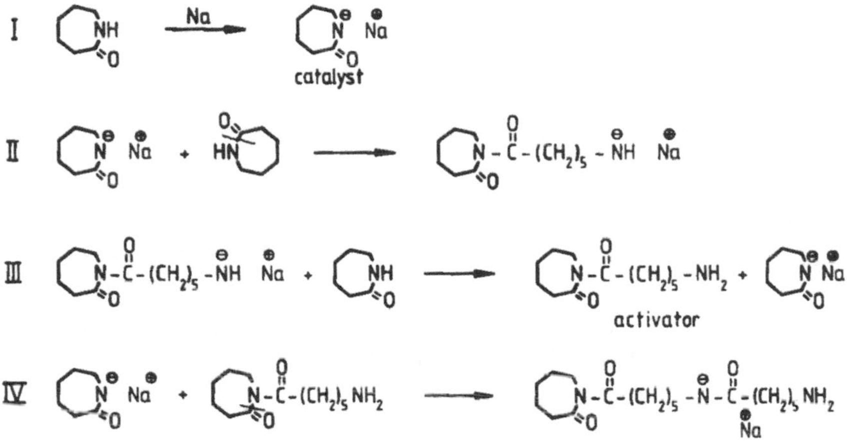

Figure 3: Anionic polymerization of caprolactam.

In accordance with the conditions, a formulation was chosen that features a high activator content and a low catalyst content. This resulted from the necessity to match the reaction to temperatures prevailing in the area ot the entry into the polimerization zone and to reduce the alkali content in the metering zone. Catalyst that is still reactive in the metering zone after conclusion of the reaction leads to an equilibrium reaction particularly at high temperatures - which entails a high residual monomer content (Fig. 4).

In addition the preliminary tests have shown that an auto-catalytic effect such as the one described by Lin and others /20/ can be made out at the low processing temperatures usual in monomer casting. However, Fig. 4 proves that there is no longer such an effect with the formulation selected for polymerization in the extruder and at the necessarily higher processing tempe-ratures.

Furthermore the stability of the formulation components in the
presence of atmospheric oxygen was tested in the course of the
preliminary tests in order to rule out the possibility of changes
in reactivity on account of the metering process.

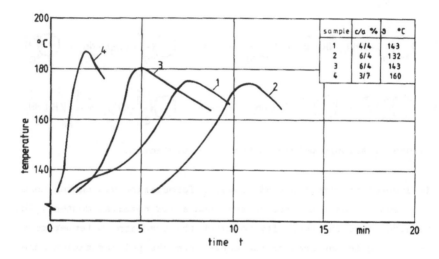

Figure 4: Progress of the activated anionic polymerization of
          polyamide 6

Polymerization in the extruder

The extruder has to take on a large number of functions in order
to perform the polymerization. These range from feeding, melting
and throughly mixing the reaction components, via the defined
start and execution of the reaction, through to the discharge of
the polymerized polyamide melt. The implementation of these
functions and the choice of operating parameters requires
consideration to a number of essential requirements which stem
from the polymerization reaction.

- Apart from the precise metering of system components
  already mentioned above, it is necessary for moisture
  to be largely eliminated if the course of the reaction
  is to be reproducible. This is particularly true for the
  catalyst, since it is deactivated by the presence of
  moisture.

- The extruder must both convey the low viscous monomer melt
  (which has the viscosity of water) and discharge the highly
  viscous polymer melt. The difference in viscosity can be
  up to five orders.

- When the caprolactam and its ingredients are being melted
  it is essential that all the components are mixed thoroughly so as to ensure that the reaction does not start in
  individual "islands" and thus cause an inhomogeneous final
  product.

- At the same time, the material must be brought to the
  reaction temperature as quickly as possible in order to
  prevent premature degradation of the additives through
  oxidation. The shear stresses can be very high during
  this phase.

- Preliminary tests /9/ have shown that the material may only
  be subjected to slight stress during the reaction. The
  mechanical and thermal loading that occurs leads to low-
  molecular products.

- The temperature dependence of the reaction kinetics makes it
  essential to have the temperature controlled in such a way
  that it is aligned as precisely as possible to the course of
  the reaction.

A screw geometry was developed and the operating parameters
determined with consideration to these requirements. Figure 5
shows the configuration of the screws as it results from
practical tests /10/.

After the feed zone with a large pitch comes a melting zone
with decreasing pitch. The monomer melt is then thoroughly
mixed by means of kneading elements.

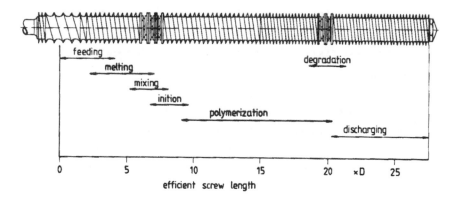

Figure 5: Screw configuration (intermeshing, co-rotating twin-
          screw extruder)

The heat necessary to start the reaction can be introduced via the
barrel into the thin film of melt that forms on the cylinder wall
on account of the kneading elements. The length of the subsequent
polymerization zone has been adapted to the reaction rate of the
system employed through practical investigations /10, 11/. Apart
from the dimensions of the extruder employed, the zone length
available for polymerization also plays a role in determining
throughput.

Before the high-molecular melt is discharged, overcoming the resistance of the die, it is possible to make provision for a kneading element which will reinforce the degradation, which has nonetheless already started, through an increased shear stress on the melt.

This defined degradation can even out molecular weight peaks. Since the kneading element constitutes a "pressure consumer" it is also necessary to build-up pressure; a back-pressure section with a maximum level of bulkfactor develops in the polymerization zone. The mean residence time thus increases due to the action of the kneading element.

The temperature profile along the length of the cylinder must be set in accordance with the different functional zones of the extruder. In the course of the process, once the reaction has started, the temperature must be adjusted to the increasing degree of polymerization and to the viscosity of the melt.

Polyamide melts produced in his way display higher molecular weights than standard polyamide 6 moulding compounds. Whilst the standard moulding compounds on the market are supplied up to mean molecular weights of some 40 000; the molecular weight range that can be reproducibly attained in activated anionically polymerized moulding compounds in the extruder is up to some 80 000. Apart from giving improved mechanical properties and good further processing properties on the part of the melt, thes high molecular weights do, however, also lead to very high melt viscosities.

The polymerizable throughput for the small laboratory extruder with an effective screw lenght of 28 x D is between 4 and 10 kg/h. An increase in throughput would be possible, within limits, by lenghtening the polymerization zone.

Process influences on the reaction

A process analysis is designed to establish the influence of
operation parameters and screw geometry on the course of the
chemical reaction in the extruder. This makes it possible to
influence the course of the reaction and the essential melt
properties, such as molecular weight, within certain limits
and to tailor these to specific applications and post
processing methods /10,11/. The influence of the operating
parameters (screw speed, throughput, temperatures) on the
mean molecular weight of the melt was thus investigated.
Solvent viscosimetry in accordance with DIN 53728 was used to
determine the average molecular weight.

In so far as polymerization has been completed in the polymerization zone, the polymerized melt subsequently undergoes
degradation on account of thermal and mechanical stressing.
This effect is reinforced by the kneading elements installed
prior to the discharge zone. Higher screw speeds for the same
throughput thus lead to lower molecular weight (Fig. 6). On the
one hand, the degradation is reinforced by the mechanical
stressing, on the other hand, however, reduced residence times in
the polymerization zone and the sensetivity of the reaction to
shear stress /9/ mean that only polymerizates with a lower
reaction rate and a lower molecular weight are formed in the
polymerization zone. At extremely low speeds, however, a
counteracting effect occurs. Long residence times and poor mixing
of the monomer melt lead to low molecular weights.

The typical effect found in monomer casting, whereby the
reacting compound undergoes direct conversion from the
monomer melt to semi-crystalline solid particles, must
be avoided. The temperature set in teh reaction zone must
therefore be above the polyamide melting point (approx.
220 °C).

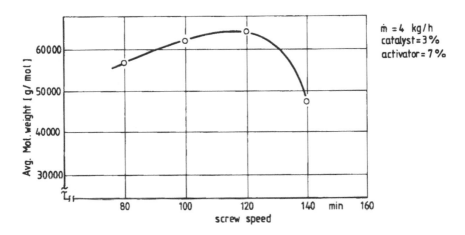

Figure 6: Influence of screw speed on the molecular weight of the
polyamide melt

Any variation in the temperature in this zone will no longer have
such a significant effect on the reaction kinetics as at the low
temperatures at which monomer casting is performed. High
polymerization temperatures lead to a reduction of the molecular
weight; the influence of the termal melt stressing predominates
(Fig. 7). This effect is highly pronounced for the monomer casting
system under investigation, since there has been no stabilisation
of the melt. Instead, the catalyst that is still active after
polymerization can even induce degradation at high temperatures.
All in all, it is possible to prove an interaction of several
mechanisms which can be influenced by the operating parameters.
Apart form the temperature and shear dependence of the reaction,
the thorough miximg of the monomer melt and also the thermally
and mechanically induced degradation effects have an essential
influence on melt quality. The residual monomer content and the
molecular weight (weight average and distribution) of the polymer
are to be regarded as factors which have a decisive influence on
melt quality.

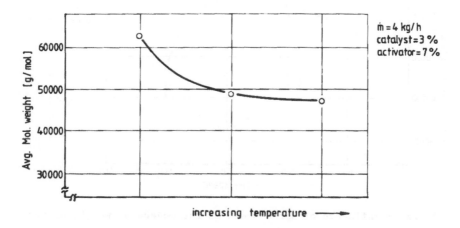

Figure 7: Influence of the temperature of the reaction on the
molecular weight of the polyamide melt.

Performance aspects of the extruder as reactor

This is the title of a paper in which Siadat, Malone and
Middleman /21/ after attempted to mathematically estimate the
course of the reaction in a single screw extruder. In terms of
its conveying characteristic the co-rotating, intermeshing twin
screw extruder is comparable with the singel screw extruder;
application of the results to the former therefore seam to be
interesting. The assumptions made by the authors are valid for
this type of twin screw extruder aswell. According to this the
reaction in the extruder can be described in a simplified way.

Using this approach, an evaluation specifically for the polymerization zone shows that, considering the high temperatures
which at least, reduce the viscosity increase caused by the
reaction, the activation energy hardly is an important factor.

Figure 8 shows the result; it is based upon measurements of the residence time in the reaction extruder (Fig. 9).

Whether at all and to what extend this straight forward approach can be applied to large machines is a question that comparative investigations will have to answer. More accurate, for example according to /21/, calculations will be necessary for higher viscosities of the reaction components at entry into the reaction zone and for reduced heat transfer conditions.

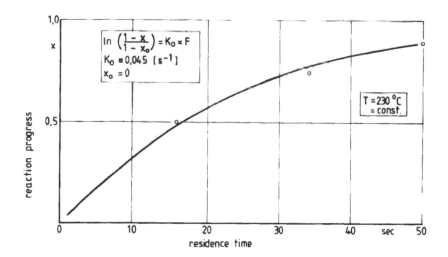

Figure 8: Effect of residence time on conversion

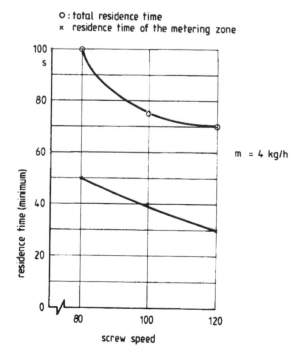

Figure 9: Influence of screw speed on the residence time

Prospect

Although there is a lot of research work to be done the importance of reactive extrusion will increase. Especially the producer
will find attractive applications like blending, grafting and
polymerization.

Using an extruder as a chemical reactor he has the opportunity to produce or to optimize polymers for special applications in a rather economical way. One example which still exists are the reactive Polystyrens from Dow Chemical. The application of these polymers in coextrusion lead to several improvements. The most important ones are the better adhesion of the different layers and the chance to save the extruder which regularly plasticats the adhesive.

The fundamentel assumption for the application of the extruder as a reactor is the comprehension of the mechanisms performing in the extruder. The chemical mechanisms of the reactions are wellknown in several cases. Thats why the most important targets of actual research work are to analyse the thermodynamical and transport mechanisms of the extruder system. This leads to rules for the design and optimization of the process.

Bibliography

/1/   Illing, G., Kunststofftechnik 7 (1968) No. 10, p. 351 - 355

/2/   Illing, G., Modern Plastics (1969) 8, p. 70 - 76

/3/   Schaaf, S., Technische Rundschau (1981), No. 7, p. 17 - 19, No. 8, p. 17 - 18

/4/   N.N., NYRIM, paper by Monsanto Europe S.A., Brüssel

/5/   Schrijver, J., RIM Nylon paper by DSM, Geleen

/6/   Kircher, K., Chemische Reaktionen bei der Kunststoff-verarbeitung, Carl Hanser Verlag, München-Wien, p. 101 - 107

/7/   Herrmann, H., Burkhardt, U., Vergleichende Analyse dicht-kämmender Gleichdrall- und Gegendrall-Doppelschnecken, Österreichische Kunststoffzeitschrift, Kunststoff-Kolloquium, Leoben, 1978

/8/   N.N., Reactions and Polymerizations using Co-rotating Intermeshing Twin-Screw Extruders paper by Werner & Pfleiderer Corporation, Ramsey, N.J.

/9 /   Ross, S., Experimental thesis at IKV, Aachen 1985

/10/   Heidemeyer, P., Diploma thesis at IKV, Aachen 1986

/11/   Singer, C., Diploma thesis at IKV, Aachen 1986

/12/   Sawada, T., Fujisawa, T., Japan Plastic Age (1983) 5, p. 36 - 40

/13/   Cremer, M., Experimental thesis at IKV, Aachen 1986

/14/   Tillmanns, P., Experimental thesis at IKV, Aachen 1986

/15/   N.N., Paper by BASF AG

/16/   Fahnler, F., Anionic, activated polymerization of caprolactam, paper by Bayer AG

/17/   Sneller, J.A., Reactive processing: New era of innovation begins in resin production, Modern Plastics Int. Aug. 1985, p. 42 - 46

/18/   Wilgolinski, L., Nangeroni, Z., The twin-screw extruder - a versatile tool for reactions and polymerizations. Advances in polymer technology. Vol 3, No. 2, p. 99 - 105

/19/   Siadat, B., et al., Some performance aspects of the extruder as a reactor, Polymer Eng. a. Sci., August 1979, Vol. 19, No 11, p. 787 - 794

/20/   Lin, D.J., et al., A kinetic study of the activated anionic polymerization of -caprolactam, Polymer Eng. a. Sci., December 1985, Vol. 25, No 18, p. 1155 - 1163

PLY MORPHOLOGY OF MOLDED SMC/RELATIONSHIP TO RHEOLOGICAL PROPERTIES AND PROCESS CONDITIONS

Jon Collister
Premix, Inc.
Research & Development Department
North Kingsville, Ohio 44068

ABSTRACT

The thickening reaction, the free radical polymerization reaction, and the mechanism of mold fill have been studied for an unsaturated polyester sheet molding compound (SMC). The thickening reaction, which is a reaction between polyester acidity and magnesium oxide, has been studied by monitoring the changes in viscoelastic response using Dynamic Mechanical and Brookfield Viscosity techniques. The extent of reaction is well correlated to the degree of neutralization of the acid groups of the polyester resin, supporting the chain extension theory of polyester thickening reactions as proposed by previous researchers. The free radical polymerization of an SMC formulation was studied by dynamic dielectric spectroscopy and differential scanning calorimetry techniques. The dielectric analysis was done on a compression molded sample in non-isothermal experiments while the DSC characterization was done at a series of isothermal temperatures which permitted calculation of the reaction rates for the SMC. Both the thickening and cure reactions have a tremendous influence on the end product performance properties of molded SMC composites; therefore, experiments were conducted to demonstrate the interaction of these two reactions with the process conditions used to mold the SMC. These experiments employed differently pigmented SMC layers in the molded charge mass which permitted easy visual inspection of the influence of process and reaction rate variables on the ply morphology of the molded SMC. These data indicate that very precise control over the viscoelastic properties and the process conditions are essential for reproducible SMC moldings.

INTRODUCTION

SMC (Sheet Molding Compound) has become an extremely important material for automotive, electrical, appliance, and corrosion applications because of attractive costs, stiffness to weight ratio, cosmetic appearance, and ability to mold complex part geometries. With this commercial acceptance has come a demand for increased understanding of the processes used to fabricate parts. The production of SMC parts involves two distinctly different chemical reactions which have major influence on the end product performance characteristics. First there is an acid/base reaction involving polyester end group acidity with alkaline earth oxides (or hydroxides). This provides an increase in viscosity and imparts elasticity to the compounded SMC, allowing flow without resin-glass separation. Secondly, there is an elevated temperature peroxide initiated free radical polymerization between the fumarate functionality of the polyester and styrene monomer to form crosslinks. This imparts high modulus and chemical resistance to the resultant molding. The end product performance of an SMC part is heavily influenced by both the thickening and polymerization reactions; the influence of each reaction must be considered in selecting process parameters since each affects ply morphology during molding.

SMC is made in relatively thin sheets (i.e. 2 mm to 4 mm) and the molds are usually quite complex; therefore, SMC is charged into the molds in multi-ply masses which require considerable flow to fill the mold. Previous studies have shown that SMC flow is completed prior to polymerization with uniform extension of the individual plies with some interply slippage possible. Little or no interply mixing is observed except in the case of severe flow restrictions. Mold geometry, charge geometry, mold closure rate, mold temperature, viscoelastic characteristics, and polymerization kinetics will all have considerable influence on final ply morphology of SMC moldings. The morphology of the plies in an SMC part will determine the end product performance, particularly in applications requiring mechanical loading in service conditions. For these reasons, it becomes extremely important to understand the interrelationship between these reactions and the process conditions used to mold the SMC.

Several researchers have performed detailed work on the polyester alkaline earth oxide reaction which contributes to the understanding of the viscosity rise in SMC. The reaction mechanism was first theorized to involve coordinate complex formation of polyester carbonyls to magnesium ions in a polyester basic salt.[1] This theory was later replaced by a chain extension theory which was supported by experiments involving controlled end group polyester resins.[2] The chain extension theory was further supported by experiments measuring the dynamic mechanical properties[3] of thickened polyester resins and by several experimental observations of actual molecular weight increases in thickened polyester resins.[4,5] The reaction as now understood is particularly exciting since a wide variety of viscoelastic responses are achievable by varying the concentration of magnesium oxide. The degree of neutralization of the polyester acidity will control the molecular weight increase of the unsaturated polyester, therefore allowing the formulator of SMC to choose the proper viscosity level for a particular application.

The free radical initiated co-polymerization of the fumarate functionality and styrene monomer is critical due to the fact that no appreciable polymerization should occur during mold flow, but a minimum residence time in the molds is required to meet productivity expectations. Differential Scanning Calorimetry (DSC) has been used by a number of workers to determine the exothermic heat of polymerization, the reaction kinetics, and cure inhibition by quinones and related inhibitors in polyester/styrene polymerization.[6] The inhibitor study demonstrated that the inhibitor provides an induction period before any polymerization occurs and that the length of the induction period was directly related to the concentration of inhibitor and the temperature. Therefore, if sufficient inhibitors are included in the SMC, mold flow is achieved during the induction period. A kinetic expression first proposed by Kamal[7] for polyester/styrene mixtures has been used to determine the kinetics of SMC polymerization. This kinetic equation is as follows:

$$\text{Equation \#1} \qquad \frac{d\alpha}{dT} = K\,\alpha^m(1-\alpha)^n$$

Isothermal polymerizations were performed on SMC at a series of temperatures which permit the calculation of reaction rates and activation energies of this polymerization.

Although extremely valuable, DSC techniques have a drawback in that very small samples of material are used and the polymerization takes place at ambient pressures. The actual practice of molding SMC is a non-isothermal process under pressure where ambient temperature molding masses are charged to an elevated temperature matched metal die where the material is heated and pressed into the shape of the mold. Heat transfer from the mold initiates the polymerization, causing an exotherm which heats the bulk of the SMC above the mold temperature. A technique to monitor the polymerization of the materials under molding conditions is DDS (Dynamic Dielectric Spectroscopy). DDS experiments have been performed on SMC at various mold temperatures to demonstrate the effect of temperature on polymerization rate. In these DDS experiments we monitored electrical damping (tan δ) as the material was molded.

With an understanding of the thickening and cure reactions, experiments were performed to investigate the influence of the viscoelastic properties and cure kinetics on the resultant ply morphology in molded SMC. Samples were molded with different pigmented colors in the various plies comprising the SMC charge. The pigmentation of the plies allows for an easy comparison of original charge configuration to the ply morphology in the resultant molding. Previous researchers[8,9] have performed a number of experiments which have indicated that the process conditions, especially mold closure rate, have a heavy influence on the resultant ply morphology. These experiments indicate that with slow closure rates the material forming the top and bottom plies of the SMC charge tend to "run" ahead of the rest of the plies. This phenomena has been theorized to be due to the lower viscosity of the surface layers due to heat transfer from the mold and, therefore, more flow is realized in these plies than in centrally located plies. Experiments with faster closure rates have indicated a uniform ply extension almost until the mold wall is reached. This condition is theorized to be due to the fact that the press closure takes place before heating of the surface plies. It has been further theorized that lower neutralization states of the SMC resin with magnesium oxide will simulate the viscoelastic properties associated with heated plies and will yield results similar to those for slow press closure rates. Furthermore, if a combination of SMC viscosities were used, it would be expected that the lower viscosity plies of an SMC charge would tend to "run" ahead of the remainder of the charge.

This phenomenon of plies "running" ahead of other plies in the charge pattern has serious implications in the US automotive industry due to the recent heavy emphasis on quality. To compete with alternate automotive body panel materials, SMC must be extremely high in quality and extremely reproducible in molded performance. Once the flow mechanisms of SMC are determined by observing the ply morphology of molded SMC parts, it becomes extremely apparent why close control of the thickening reaction, polymerization reaction, and press closure conditions are required.

The object of this work is to present an overview of the analytical techniques used to characterize the thickening and polymerization reaction; and then qualitatively to demonstrate how these two reactions influence the ply morphology of SMC. The information can be used to explain observations of SMC molding performance in production applications. It is intended that an understanding of the critical influence of these reactions on SMC ply morphology will create more emphasis on control and allow exploitation of the wide latitude of properties achievable with SMC.

EXPERIMENTAL MATERIALS

The SMCs used in this study were typical of the SMCs which have found usefulness in the automotive industry. The following formulation was used in this study. Variations from the formula were made by altering the pigment type to permit visualization of the various SMC plies in resultant moldings, and altering the thickener concentrations to achieve desired SMC viscosity.

Propylene Glycol Maleate Polyester in 35% Styrene	60 phr
Polyvinyl Acetate (acid functionalized) in 60% Styrene	40 phr
t-Butyl Perbenzoate	1 phr
Zinc Stearate	4 phr
Pigment (varied)	1 phr
Calcium Carbonate	175 phr
Thickener	Varied to achieve desired viscosity
2.54 cm glass fibers	28% by weight

All materials used are commercially available products which were used as supplied with no modifications. The pigments used to impart color to the various plies were commercial proprietary pigment dispersions in white and red. They were purchased and used in the SMCs without analysis of composition.

The SMCs used in this study were compounded on a laboratory SMC line which made SMC approximately 60 cm in width. All materials were manufactured and then allowed to thicken for seven days prior to any molding evaluations.

The polyester resin used in the Dynamic Mechanical measurements of "resin only" experiments was a commercially available polymer condensed from propylene glycol, maleic anhydride, and isophthalic acid. The magnesium oxide was a commercial grade of 85% purity.

EXPERIMENTAL PROCEDURES

A. Dynamic Mechanical Testing - Dynamic mechanical evaluations were conducted using a Rheometrics Thermal Mechanical Spectrometer equipped with parallel plate testing fixtures of 25 mm and 50 mm diameters.

B. Brookfield Viscosities - Brookfield viscosities were measured with a Brookfield HBT-5X viscometer equipped with standard t-bar spindles, all measurements were made with the TF spindle at 5 rpm to eliminate shear rate effects.

C. DSC - Experiments were performed with a Perkin-Elmer DSC-2 interfaced to an IBM PC-XT® for data collection.

D. DDS - Dielectric analysis was performed with a Tetrahedron Audrey® Dielectric Analyzer equipped with a parallel plate test cell. The electrodes were mounted in the mold surfaces of a 15.25 cm X 15.25 cm matched metal die in a Laboratory press.

E. Ply Morphology - Morphology experiments were conducted on a 30.5 cm X 30.5 cm matched metal die mold. Cross-sectional views were obtained by sawing the molding and polishing to enhance the visibility. "Short shot" experiments were performed by placing stops on the mold to the designated thickness.

RESULTS AND DISCUSSION

Thickening Reaction

The mechanism of polyester thickening which is theorized to involve a molecular weight increase in the polyester through formation of magnesium/polyester salts is consistent with data which has been generated regarding the dynamic mechanical properties of thickened polyester resins.[3] Burns,[2] et al, showed that by control of end group functionality of the polyester, a wide variety of thickening levels are achievable with the same molecular weight resin. These data indicate that doubly acid terminated polyester species are those primarily responsible for the large viscosity increase in polyester resins. Polyesters as commercially produced are variously terminated; some species exist as doubly hydroxyl terminated species, some species are both acid and hydroxyl terminated, and other species are doubly acid terminated. This termination population indicates that some of the polyester molecules will not be able to enter into the polyester magnesium salt formation due to termination with hydroxyls. Previous investigations[3] of thickened polyester resins indicate that the viscoelastic response of these materials is very similar to the response exhibited by low molecular weight/high molecular weight blends of thermoplastic materials. Furthermore the blend of high molecular weight and low molecular weight components exhibit viscoelastic responses which are dominated by the high molecular weight portions.[10] This is demonstrated in Figure #1 and #2, which display G' versus frequency for a polyester and a polystyrene material, respectively, and Figure #3 and #4 which display G" versus frequency for the same samples. Figure #1 and #3 show data for a magnesium oxide thickened polyester resin at various times after addition of magnesium oxide, while Figure #2 and #4 display data for polystyrene blends of low molecular weight and high molecular weight materials. The data are very similar, which supports the concept that the small proportion of doubly acid terminated species in the polyester material can be primarily responsible for the viscosity increases observed in polyester resins.

Another finding of this previous work[3] was that variations in MgO concentration produced equilibrium viscoelastic responses analogous to the progression of viscoelastic responses associated with a single high concentration over time. It requires approximately one week to reach the equilibrium

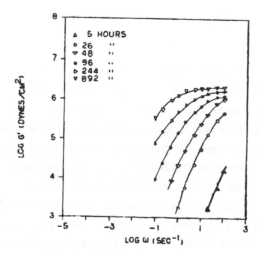

Fig. 1. G' vs. ω for a thickened polyester resin at various times after addition of MgO.

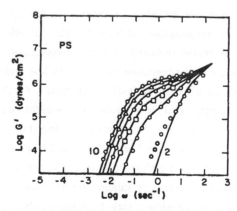

Fig. 2. G' vs. ω for polystyrene blends. The outer curves are for pure components. The other curves represent 20/80, 40/60, 60/40, and 80/20 blends. Data from Soong, Shen, and Hong[9].

neutralization value at which time the viscosity, as measured by Brookfield, reaches a plateau. The plateau viscosity is where the polyester/magnesium salt is in equilibrium with the reactants. This phenomenon allows tailoring of individual viscosity states for individual applications in end use molding since MgO concentrations can be manipulated to control the equilibrium viscoelastic response. A detailed MgO concentration study has been performed and it has been found that the polyester magnesium oxide interaction is

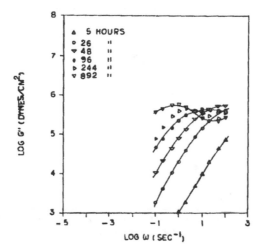

Fig. 3. G' vs. ω for a thickened polyester resin at various times after addition of MgO.

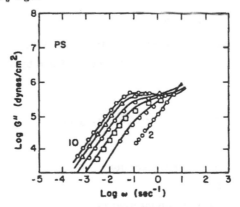

Fig. 4. G" vs. ω for polystyrene blends. The outer curves are for pure components. The other curves represent 20/80, 40/60, 60/40, and 80/20 blends. Data from Soong, Shen, and Hong[9].

quantitative. In Figure #5 data is presented regarding viscosity versus the molar ratio of magnesium oxide to polyester acidity. In this plot we see that the maximum viscosity is approached at a 1:1 molar ratio.

Referring to Figure #5, it is easy to see how a formulator using different amounts of magnesium oxide has a wide variety of viscosity states available in one SMC formulation. Judicious use of the thickening agents in polyester SMC allows custom formulating for almost any molding geometry and process conditions.

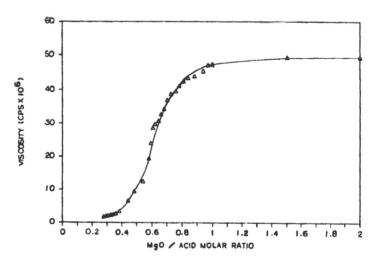

Fig. 5.   72 Hour Viscosity vs. MgO/Acid Molar Ratio for SMC.

Polymerization Reaction Kinetics

Previous studies have indicated that the cure kinetics as measured by Differential Scanning Calorimetery can be well defined by a kinetic expression proposed by Kamal[7]:

$$\frac{d\alpha}{dt} = K \, \alpha^m \, (1-\alpha)^n$$

Where:

α	=	Monomer Conversion
t	=	Time
K	=	Rate of polymerization
m	=	Constant = 0.5
n	=	Constant = 1.5

The integration of DSC exotherms is used to give the information relating to the degree of monomer conversion (α). Previous studies in polyester kinetics have primarily centered around the pure polyester resin-styrene mixture with various peroxides. Koboto[11] did work regarding pressure DSC of SMC formulations which indicated that results similar to pure resins are obtained. In this study, we have extended the work to fully formulated SMCs and have verified the fact that the kinetic expression proposed by Kamal is adequate

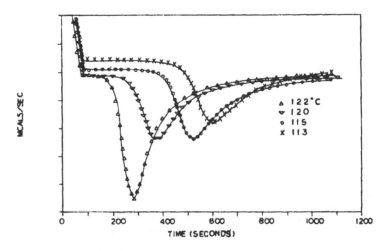

Fig. 6. Isothermal DSC Traces at Various Temperatures.

to describe the polymerization kinetics of SMC materials.

Iosthermal heats of polymerization $(\Delta H)_I$ were determined for SMC pastes at 113°C, 115°C, 120°C, and 122°C. Additionally, residual heats of polymerization $(\Delta H)_R$ were determined in a temperature ramping mode to 220°C. This residual $(\Delta H)_{RES}$ was added to the isothermal polymerization to permit the calculation of conversion (α) data as a function of time where:

$$\alpha = \frac{(\Delta H)_I}{(\Delta H)_I + (\Delta H)_{RES}}$$

Where:

α	=	Monomer Conversion
$(\Delta H)_I$	=	Isothermal Heat of Polymerization
$(\Delta H)_{RES}$	=	Residual non-isothermal Heat of Polymerization

The first derivative of the α versus time plot was computed and then plotted as a function of $\alpha^m(1-\alpha)^n$ to determine the reaction rate of the material. The reaction rates and regression coefficient were determined at each isothermal temperature and are presented in Table I. Excellent linearity of $d\alpha/dt$ versus $\alpha^m(1-\alpha)^n$ was observed. Isothermal ΔH_I diagrams are presented in Figure #6, which clearly show that the induction period before polymerization is directly related to temperature.

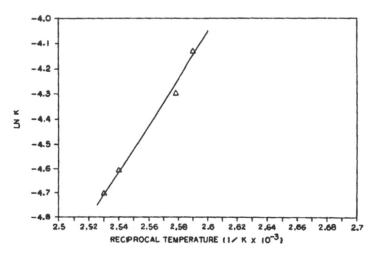

Fig. 7. Logarithm of Polymerization Rate vs. Reciprocal Temperature for Isothermal DSC Data.

The logarithms of the calculated rates were then plotted as functions of 1/T to verify Arrhenius type temperature dependence and to determine the activation energy of this reaction. We determined from this plot (which is included as Figure #7) that the activation energy of this reaction was 18.4 kcal/mole; this agrees well with previous data for polyester polymerizations. We also averaged 27.3 cal/gram as a ΔH_T value which also agrees well (after adjustment for inert ingredients) to work performed by previous researchers. Extrapolation to zero inert concentration yields 82.6 cal/gram normalized for the base resin material, while previous reported data predicts 70 to 75 cal/gram. The particular resin used in this formulation contains more styrene monomer than those resins analyzed by Kamal and other workers; therefore, we do expect to see higher heats of polymerization due to the higher concentration of unsaturated species.

The actual molding of SMC is of course a nonisothermal process since the molding masses are charged at ambient temperature into an elevated temperature matched metal die. The temperature profile of an SMC after charging to the mold depends upon the distance from the mold wall and the heat transfer characteristics of the SMC. Presented in Figure #8 are two plots of temperature versus time for an SMC material which has been placed into a compression mold. The temperature profile A is for a non-initiated material which shows no polymerization exotherm and, therefore, the

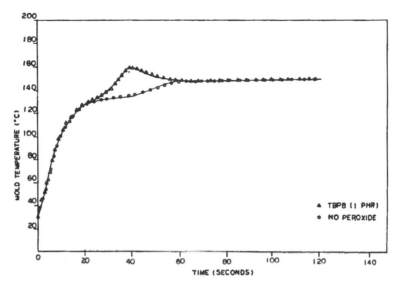

Fig. 8. Temperature vs. Time plots of SMC in a mold at 150°C. Profile A is for a non-initiated SMC while Profile B is for an SMC with 1 phr TBPB.

temperature rise is simply due to heat transfer. Temperature profile B is from an actual SMC which contains peroxide and shows the corresponding exothermic activity after 20 seconds. A gradient of temperature profiles will be generated depending on the distance from the mold wall and the distance down a flow channel. Due to these conditions, the extrapolation of DSC polymerization kinetics information to actual molding practice is extremely difficult and a variety of estimations would have to be made to make predictions on an actual production part. Fortunately, a technique exists (DDS) which provides very practical polymerization information on the material in an actual compression mold. In Figure #9, tan δ information is presented as a function of time for an SMC molded at 138°C, 150°C, 160°C, 171°C, and 182°C. The initial portion of the curve exhibits an increase in tan δ due to a viscosity drop associated with material heating. As the material starts to polymerize, a rapid decline of tan δ is observed due to the restriction in dipole mobility due to cross-linking. Although this data is not easily converted to precise kinetic data since it is not performed under isothermal conditions, it offers a very practical way of assessing the time limitations for mold flow. After the maximum point in tan δ is reached, the material

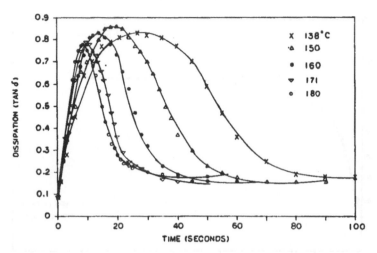

Fig. 9. Tan δ vs. Time Plots for DDS Experiments at Various Mold Temperatures.

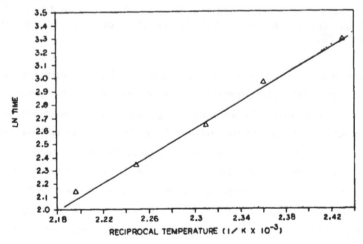

Fig 10. Logarith of Time (to Tan δ max) vs. Reciprocal Temperature for DDS Experiments.

gels very rapidly and further mold flow is impossible after tan δ_{max}. This technique provides a very convenient way of assessing the proper initiator concentrations, initiator types, and appropriate mold temperature.

The nonisothermal nature of SMC molding complicates calculation of precise reaction rate data but it allows other calculations since tan δ_{max} is a distinguishing point in this polymerization; that point in time can be used

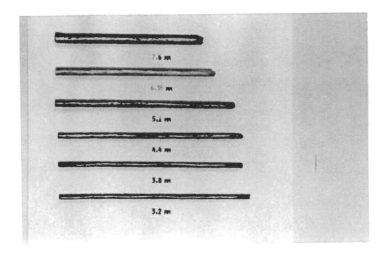

Fig. 11. Progressive Ply Morphology of an SMC Molding during closure.

to determine the activation energies of the reaction. Figure #10 was prepared by plotting the logarithm of time to achieve tan δ_{max} as a function of $1/T$. The straight line exhibited in this plot again indicates the Arrhenius dependence of this polymerization reaction on temperature. The correlation coefficient for this plot is 0.996, indicating a good fit to the Arrhenius assumption. This technique does, however, give a much lower activation energy (10.1 kcal/mole) than the DSC experiments. This difference in activation energy may be a consequence of the fact that the time to tan δ_{max} is dependent on the decomposition reaction of the peroxide and interaction between these free radicals and the included inhibitors whose kinetics have not been determined.

Molded Ply Morphology

The work performed in this phase of our evaluation confirmed previous work reported in the literature.[8,9] Figure #11 shows the progressive ply morphology of SMC at various times during mold closure. It can easily be seen that the outer plies "run" out to the front of the flow so that the edges of the molding are comprised of material which have been in close contact with the mold walls. The "running" of plies at slow closure rates was observed exactly as depicted by Barone, while at fast closure rates a more uniform

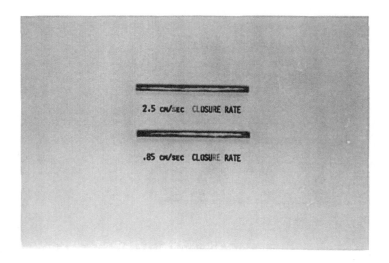

Fig. 12. Relationship of Mold Closure Rate and Molded SMC Ply Morphology.

flow of the plies was observed. These effects are presented in Figure #12 for closure rates at 2.5 cm/sec. and 0.85 cm/sec. The slower closure rate caused differential flow velocity between the surface layers and internal plies, which is theorized to be due to the heating of the plies adjacent to the mold surfaces. At the slower rates the temperature differential between surface plies and inner plies is greater, thus lowering the viscosity of the plies allowing this "running" effect. In this regard, evidence was generated indicating that a long residence on the mold before initiation of mold closure causes "running" of the bottom ply which creates a much different fill pattern in the periphery of the parts. Figure #13 shows the ply morphology of two SMC moldings with different residence time on the mold prior to mold closure. This figure confirms the presence of an induction period before polymerization occurs. The longer residence time example displays no evidence of gellation, since gellation would have prevented flow in the outer layers. To further confirm the theory of higher heating rates causing the "running" of plies, an SMC formulation was made with reduced magnesium oxide concentration to provide a lower viscosity SMC. In these experiments, lower viscosity (17×10^6 cps) outer plies were used while the internal plies were at higher

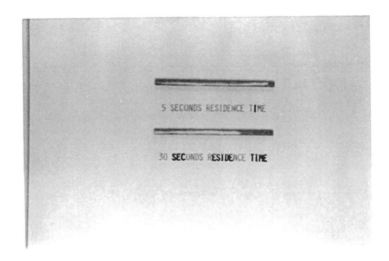

Fig. 13. Relationship of Residence Time on Mold (Prior to Mold Closure) and SMC Ply Morphology.

viscosity (34 X 10^6 cps). The lower viscosity outer plies simulate the viscoelastic condition that occurs when the outer plies are substantially heated by contact with the hot mold surface. These results, as shown in Figure #14, show exactly the same effect as the material which had been in contact with the mold surface for longer periods of time. We can therefore qualitatively say that the model of the outer ply heating and allowing "running" of the material during slower close rates is a valid theoretical approach. To further demonstrate this effect, an SMC molding in which a low viscosity ply which was placed in the middle of the charge is given in Figure #15. As can be seen from the photograph, the ply which was placed in the middle "ran" out in front of the flow front which indicates that maximum ply deformation will occur in the lowest viscosity ply, regardless of whether the lowest viscosity is achieved by heat transfer effects or thickening differences.

Another series of experiments were run to determine the charge pattern influence on the ply morphology. In these experiments, panels were molded with 11%, 25%, and 44% mold coverage. (To achieve the same mass, more plies were molded for the lower coverage charge patterns.) As can be easily seen from Figure #16, the 44% mold coverage material produces a flow pattern

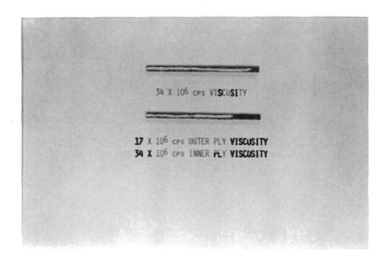

Fig. 14.　SMC Ply Morphology with Varied Viscosity Outer Plies.

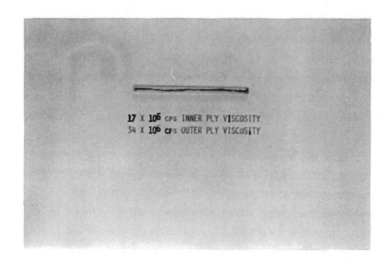

Fig. 15.　SMC Ply Morphology with Low Viscosity Inner Ply.

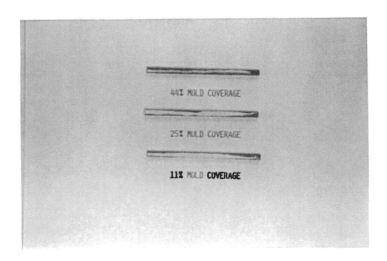

Fig. 16. Relationship of SMC Ply Morphology to Charge Mold Coverage.

with the entire surface of the panel being created by biaxial extension of
the outer charge layers. The 25% and 11% mold coverage panels show
considerable flow disturbance of the upper layer indicating that the inner
layers in certain areas become some of the molded surface. This "flow through"
to the surface by the inner plies can cause disruptions in the surface appearance
of the material leading to reduced quality molded parts.

Experiments were conducted at fast closure rates at 140°C, 150°C, 160°C,
and 170°C to demonstrate the effect of mold temperature The results of
these experiments are presented in Figure #17. The mold temperature does
not affect the morphology of the molded plies to any great degree. It appears
that at these closure rates the heating of the outer SMC plies is not
substantially affected by the incremental increase in mold temperature.
Additionally, it again appears that even at higher mold temperatures the
polymerization reaction does not become a dominant factor during the flow
period.

CONCLUSIONS

SMC is a material which depends on two independent reactions to achieve
its desirable end use properties. This work has indicated that the
polyester-magnesium oxide reaction, which is theorized to be a molecular

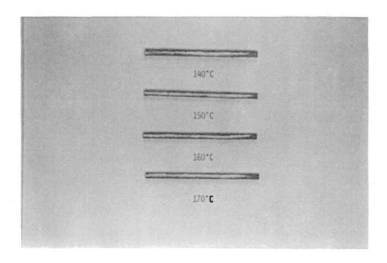

*Fig.17.   Relationship of SMC Ply Morphology to Mold Temperature
(Fast Closure Rates).*

weight increase, is quantitative with respect to the acidity and magnesium
oxide. Therefore, viscosity of SMC can be manipulated by changing polyester
acid functionality or changing the concentration of magnesium oxide. The
viscosity was demonstrated to have a large influence on the flow behavior
of SMC charges with low viscosity plies demonstrating much more flow than
higher viscosity plies. This ability to manipulate the viscosity of SMC permits
specialty formulations to be prepared to suit a variety of mold geometries
and process conditions.

The second reaction is an elevated temperature peroxide initiated free
radical polymerization. Control of this reaction is essential to permit mold
fill time and yield appropriate productivities. This reaction of SMC was studied
by DDS and DSC techniques. The isothermal DSC polymerization was shown
to behave according to a kinetic expression which was proposed for pure resin.
Both the isothermal DSC and non-isothermal DDS data were shown to exhibit
Arhennius temperature dependence. The influence of the polymerization
on the ply morphology was minimal under the conditions investigated,
supporting the concept that initiator technology has developed to minimize
polymerization until just after flow is completed.

TABLE I

Isothermal DSC Data

T (°C)	Reaction Rate (k) (sec^{-1})	ΔH (cal/gr)	Regression Coefficient*
113	9.060 X 10^{-3}	27.3	0.985
115	9.990 X 10^{-3}	28.3	0.960
120	1.243 X 10^{-2}	26.3	0.983
122	1.600 X 10^{-2}	27.2	0.980

*d/dt vs. $\alpha^m(1-\alpha)^n$

REFERENCES

1. Szmercsanyi, V.I., Kunststoffe Bel, "Interaction Between Unsaturated Polyester Resins and Fillers"; 58, 1968, 907-912.

2. Burns, R., Lynskey, B.M., and Gandhi, K.S., Research Projects in Reinforced Plastics, "Variability in Sheet Molding Compound"; RPG Fourth Conference, 1976.

3. Gruskiewicz, M., Collister, J., Polymer Composites, 3, 6 (1982).

4. Ludwig, C., Collister, J., Proc. XXXIV; Annual SPI Conference, Reinforced Plastics/Composites; 1979, Section 24-C.

5. R.S. Rapp, Proc. XXXIV Annual SPI Conference, Reinforced Plastics/Composites Institute; 1979, Section 16-F.

6. Proffitt, D., Proc. XXXVII Annual SPI Conference, Reinforced Plastics/Composites Institute; 1982, Section 22D.

7. Kamal, M.R., Sourour, S., Polymer Engineering and Science, 13, 59, (1973).

8. Barone, M.R. Caulk, D.A., Polymer Composites, 6, 105 (1985).

9. Kanagendra, M., Fisher, B.C., Proc. XLI Annual SPI Conference, Plastics/Composites Institute, 1985, Section 16-C.

10. Soong, D., Shen, M., Hong, S.D., Journal of Rheology, Volume 23, No. 3; Pages 301-322.

11. Kubota, H., Proc. XXX Annual SPI Conference, Reinforced Plastics/Composites Institute; 1975, Section 1-F.

Viscoelastic Properties of Thermoplastic Ethylene-Propylene Elastomers Dynamically Cross-linked

L. Corbelli, S. Giovanardi, E. Martini

1) FORWARD

Thermoplastic elastomers are rubbery materials which can be processed by equipment usually associated with thermoplastic resins. There is an obvious growing interest in products that can be turned directly into finished goods in striking contrast with the lengthy three-step process required by conventional elastomers: compounding, shaping and curing. It is, therefore, no surprise that in the past twenty years or so there has been an upsurge in thermoplastic elastomers, either block polymers or polymer blends which, in turn, are either uncured or dynamically cured, all trying to fit themselves to particular markets which were as yet to be defined.

Order has since come out of chaos and the host of products put on the market have gradually found their application areas. Among these, the dynamically cured PP/EPDM blends have materialized into a number of interesting, large volume products thanks to a very good cost/property compromise. Now that the situation has settled and the market is mature, time is ripe for a further study of the effect the EPDMs used have on the visco-elastic properties of the final product. As the characteristics achieved by appropriate PP/EPDM ratios span a very wide range, from stiff to extremely soft and rubbery (see Fig. 1 and 2 for a general map of properties), this work was limited to those low hardness, highly elastomeric thermoplastic vulcanizates containing comparatively small amounts of polypropylene (23% volume).

For this class of products it was felt that a correlation between the properties of the EPDM used, in terms of composition and unsaturation level, and the behaviour of the thermoplastic vulcanizate would most likely be found. However, it is now believed that the conclusions drawn in this work can likewise be extrapolated to stiffer products and that the data presented will provide some useful insights into the intimate structure of the system.

2) EXPERIMENTAL

2.1. Thermoplastic vulcanizate preparation

The general recipe used in this work is presented in Table 1 along with the curing system examined. The EPDM types cover the composition and ENB ranges indicated in

PROPERTIES OF PP/EPDM

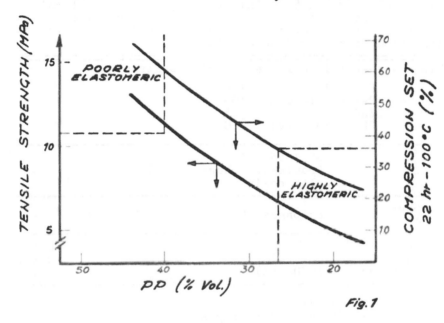

Fig. 1

PROPERTIES OF PP/EPDM

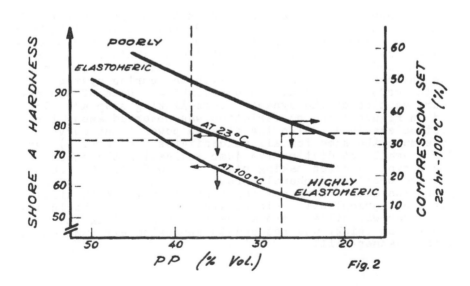

Fig. 2

in Figure 3 whereas in Table 2 the vulcanizate design codes are reported. All polymers were made in laboratory according to the well known Dutral suspension process (1) at approximately the same intrinsic viscosity (approximately $[\eta]_T^{35°C} = 4.5$ so that after paraffinic oil extension by means of a week's contact between polymer crumbs and oil at room temperature, the Mooney Viscosity values indicated in Figure 3 were obtained.

Table 1: General Recipe and Curing Systems

EPDM (1)		100		
Paraffinic oil		100		
ZnO		5		
PP - x30 G (2)		60		
Resin	SP 1055 (3) Brominated phenolic resin	10	-	-
Sulphur Donor	Sulphasan R (3)	-	4	-
Sulphur plus Accelerator	S MBT (4) TMTDS (4)	-	-	0.5 0.8 1.2

(1) Dutral S.p.A.
(2) Himont Italia S.p.A.
(3) Monsanto
(4) ACNA

The thermoplastics were prepared in two steps. First oil-extended rubber was blended with PP and ZnO in a 1.8 liter internal mixer operating at 60 r.p.m. for 8 minutes reaching a final blend temperature of 180°C. This blend was then milled into a sheet, cooled and fed once more into the internal mixer, this time with the curing system so as to perform the dynamic cross-linking. Data of temperature vs time of the dynamic cross-linking step are plotted in Figure 4 and no variation was observed among the systems examined. Also the time vs power input curves were much the same for all vulcanizates (see Fig. 5) but for a small yet distinctive jump indicating the onset of the curing reaction taking place at 150°C, 170°C, 140°C for the brominated phenolic resin, sulphur donor and sulphur plus accelerator systems respectively. When still hot the thermoplastic vulcanizate removed from the internal mixer was milled into a sheet. Upon cooling the sheet, about 3 mm thick, was reduced to approximately 3 mm chips with a hammer-mill.

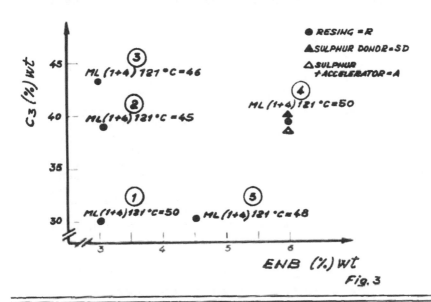

Fig. 3

Table 2: PP/EPDM Vulcanizate Designation Code

EPDM type	C_3 (%wt)	ML(1+4)121 oil extended polymer	ENB (%wt)	Curing System	Code
Experimental	28	50	3	Brominated phenolic resin	1 - R
"	38	45	3	"	2 - R
"	44	46	3	"	3 - R
Experimental	38	50	6	Brominated phenolic resin	4 - R
				Sulphur + accelerator	4 - A
				Sulphur donor	4 - SD
Experimental	28	48	45	Brominated phenolic resin	5 - R

2.2. Thermoplastic vulcanizate properties

Mechanical properties were determined according to ASTM
specifications on samples cut from a 400x400x2 mm platen
press sheet prepared as follows:
- fill the mold evenly with 5% more chips than required
by mold volume;
- pre-heat at 200°C for 20 minutes without pressure;
- press at 200°C for 5 minutes at 4 MPa;
- cool down to room temperature under pressure.

A platen press was preferred over injection molding after a preliminary investigation carried out with 5-R vulcanizates. It was found that specimens cut from platen
press sheets were isotropic and their properties were
much the same over a wide range of pressing conditions
whereas injection molding gave anisotropic specimens whose
properties were strongly dependent on injection conditions
and, at their best, only somewhat superior to those presented by platen – press prepared specimens. Hence, to a-
void the need to optimize injection molding conditions

DYNAMIC CROSS-LINKING
TEMPERATURE VS. TIME

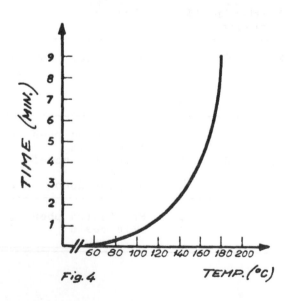

Fig. 4   TEMP. (°C)

for all the vulcanizates studied and to specify specimen cutting directions and, since the main object of this work was to assess and compare the final properties of thermoplastic vulcanizates made with different types of EPDMs, our evaluation was limited to platen press specimens.

2.3. Iso-cross-linked pure gum EPDMs

Pure gum EPDMs were prepared using the curative recipes shown in Table 3 where, for the sake of comparison, those used for thermoplastic vulcanizates are also presented. Such recipes, as will be commented upon in the section entitled "Summary of previous work and literature review" impart to pure gum EPDMs the same cross-linking density experienced by the dispersed rubber phase present in the thermoplastic vulcanizate made with the same type of EPDM and corresponding curing system. Mechanical properties were determined according to ASTM specifications on sam-

DYNAMIC CROSS-LINKING TIME VS. POWER IMPUT

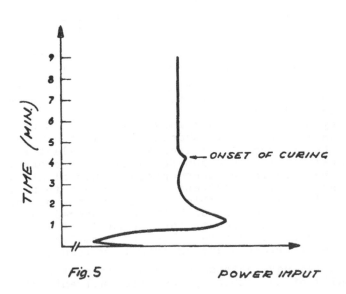

Fig. 5   POWER IMPUT

Table 3: Curative Recipes

Curing Agent	Thermoplastic Vulcanizate	Iso-crosslinked EPDM
	pph (oil+EPDM+PP+ZnO)	pph (oil+EPDM+ZnO)
SP 1055	10	6
Sulphasan R	4	2.4
Sulphur	0.5	0.3
MBT	0.8	0.5
TMTDS	1.2	0.7

ples cut from 120x120x2 mm platen press sheets cured for
15 minutes at 160°C; 170°C, 155°C with brominated phenolic resin, sulphur donor and sulphur plus accelerator respectively.

3) SUMMARY OF PREVIOUS WORK AND LITERATURE REVIEW
Before presenting and discussing the results of this work
it is worth reviewing the present state of knowledge on
thermoplastic vulcanizates (2), (3), (4), (5), (6), (7)
as well as mentioning some data obtained in our previous
works (8), (9).

For a dynamically cured PP/EPDM thermoplastic vulcanizate
it is generally agreed that:
- PP is the hard continuous phase;
- EPDM is the soft dispersed phase;
- There must be some sort of physical interaction between
the soft and hard phase "if useful properties are to be
realized" (4);
- For the best final properties the rubber particle must
be less than 20 , preferably on the order of $1-5\mu$

To this basic background, experimental results achieved
in our previous work are to be added. These can be summarized as follows:
1) The key to obtaining a highly elastomeric thermoplastic vulcanizate with low hardness is oil at a weight e-
qual to that of the high molecular weight elastomer.
2) To impart to the pure gum EPDMs the same cross-linking
density observed in the thermoplastic vulcanizate dispersed rubber phase, vulcanization recipes are to be adjusted (9) as indicated in Table 3. The cutting of curative
level by 40% in the case of EPDMs as compared to thermoplastics, not only accounts for the hard/soft volume ratio but also for an apparent loss of curative or lack of
efficiency experienced during dynamic vulcanization.

These results were determined by measuring the cross-linking density of the rubber phase after solvent extraction

of PP on non-oil extended thermoplastic vulcanizates. This
is because the method used, described by E. F. Cluff and
others (10) involving the measurement of equilibrium compression modulus at very small deformation on swollen
samples, cannot be applied to oil extended systems. The
resulting cross-linking density was then compared with
the cross-linking density of the non oil extended pure
gum cured with various amounts of the curing system in
question.

Finally, the following two assumptions which could not be
verified but which seem resonable must be mentioned:
1) With the curing system investigated, PP does not undergo any chemical modification during processing;
2) Most of the oil remains in the soft rubber phase during dynamic curing and even more so on cooling of the
thermoplastic. The literature and experimental background
as well as the assumptions behind our work are summarized
in Table 4.

Table 4: Literature, Experimental Background and Assumptions

Literature Background	PP is the hard continuous phase EPDM is the soft dispersed phase
	There are physical interactions between the soft and hard phases
	Particle sizes must be small
Experimental Background	Oil in rubber is the key compo- nent for low hardness thermo- plastics
	Partition coefficient of cura- tives depends on hard/soft volumes ratio
Assumptions	PP does not undergo chemical and physical modification during dy- namic vulcanization of the soft phase
	Most oil keeps in the soft rubber phase during dynamic vulcaniza- tion and more so on thermoplastic cooling

4) EXPERIMENTAL RESULTS AND DISCUSSION

The viscoelastic properties of thermoplastic vulcanizates and iso-cross-linked EPDMs are presented in Table 5. Data show that no matter what type of curing system and EPDM type are used the thermoplastic tensile properties, i.e. Young modulus, tensile, hardness and elongation are much the same whereas iso-cross-linked pure gums feature quite large variations in tensile properties depending on the EPDM used; the best results being shown by the low C_3 EPDM compositions 1 and 5.

On the other hand, elastic properties such as tension set, retraction test (TR-50), stored energy after 1st cycle (as defined in Fig. 6) are strongly dependent on EPDM type both in thermoplastic vulcanizate and iso-cross-linked gums; the best results being shown by the high C_3 compositions 3 and 4. It is worth mentioning that the ENB

STORED ENERGY 1st CYCLE
STRAIN AND RECOVERY RATE 30% MINUTE

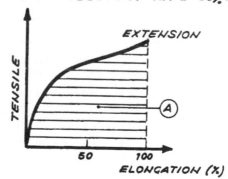

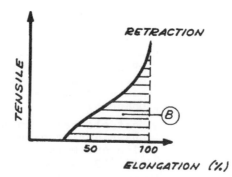

$$STORED\ ENERGY = \frac{B}{A} \cdot 100$$

Fig.6

Table 5: Properties of Thermoplastic Vulcanizates and Iso-crosslinked EPDM

		TENSILE PROPERTIES				ELASTIC PROPERTIES		
	Code	Young Modulus (MPa)	Tensile (MPa)	Elongation (%)	Shore A Hardness	Tension Set 200% 23°C (%)	TR-50 (°C)	Stored Energy 1st Cycle (%)
Thermoplastic	1 R	20.6	7.0	350	72	34	0	20
	2 R	14.6	6.8	340	70	17	-16	25
	3 R	15.1	6.9	360	71	16	-32	35
	4 R	14.5	6.5	370	72	14	-34	35
	4 A	14.2	6.8	360	70	15	-30	35
	4 SD	14.6	7.0	370	71	15	-32	40
	5 R	17.7	7.2	350	71	22	-12	25
Iso-crosslinked EPDM	1 R	6	5.5	450	30	12	- 8	50
	2 R	4	4.0	550	26	6	-22	60
	3 R	2	3.5	530	20	4	-46	65
	4 R	3	4.0	500	18	4	-44	70
	4 A	3	3.8	500	20	4	-45	65
	4 SD	3	4.1	460	18	4	-43	65
	5 R	5	6.1	550	35	10	-15	55

level has a quite negligible effect in passing from 3% of composition 3 to 6% of composition 4, possibly because in these cases the minimum level of cross-linking has already been reached. After this minimum has been reached the elastic properties, and in particular the tension set, level off (6). Therefore, it can be concluded that thermoplastic vulcanizate tensile properties must be governed by the PP hard continuous phase with negligible contribution from the rubber. On the other hand, elastic properties are mainly determined by the soft dispersed phase and the elastomer composition is the main factor in affecting elastic properties, whereas the ENB level plays a minor role.

The latter argument is further supported by an evaluation of thermoplastic vulcanizate elastic properties under various temperature and frequency conditions. Testing are most severe following this decreasing order: Ball drop (-50°C, high frequency), > Retraction test (low temperature, low frequency); > 23°C tension and compression test,> 100°C tension and compression tests.

The data shown in Table 6 indicate that the more severe the testing conditions the larger the differences among EPDM types, hence the greater the advantage in selecting the suitable high C_3 EPDM elastomer. Passing from 23 to 100°C, differences tend to be cancelled out and, therefore, any type of EPDM can be used in the preparation of thermoplastic vulcanizates intended for end products with service temperatures above 50°C or so. The reasons behind this behaviour, which hold true for both PP/EPDM vulcanizates and conventional EPDM vulcanizates, have been dealt with in some detail elsewhere (11).

It will suffice here to say that EPDM composition, composition distribution, molecular weight distribution and ethylene sequences determine the flexibility of the molecular chain as well as its tendency toward crystallization and hence the viscoelastic properties of the elastomer and compositions made thereof. In the case of thermoplastic vulcanizates there must be some sort of physical interaction between the PP and EPDM domains which, during the deformation process, allows for the storage and building up of energy within the soft phase; energy that will be consumed in the, at least partial, recovery of the deformation experienced by the PP phase. Had PP been taken alone it would not have been possible to recover this deformation as can be seen from data presented in Figure 7. The difficult points to reconcile are the ways PP domains can:
- support most of the external stress during deformation;
- transfer the deformation to the rubber phases for energy to be stored;
- recover the strain under the effect of an internal stress exerted by the soft domains.

Table 6: Thermoplastic Vulcanizate Elastic Properties vs Temperature

Code	Room Temperature			Low Temperature		High Temperature 100°C	
	Tension set 20%	Tension set 100%	Compression set 22 h %	TR-50 (°C)	Ball drop Energy to break at -50°C ( ) (J)	Tension set 100%	Compression set 22 h %
1 R	34	13	28	0	0.4	4	20
2 R	17	6	16	-16	21	4	18
3 R	16	6	16	-32	19.7	4	19
4 R	14	5	13	-34	18.7	4	18
5 R	22	10	23	-12	0.2	4	19

( ) - Himont Method DIMP MA N° 17238

EXTENSION AND RETRACTION CURVES
STRAIN AND RECOVERY RATE 30%/MINUTE

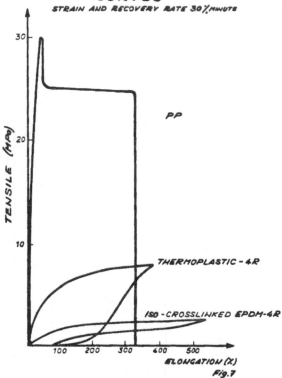

Fig.7

"PSEUDO - MULLINS EFFECT" ON THERMOPLASTIC VULCANIZATE
STRAIN AND RECOVERY RATE 30%/MINUTE

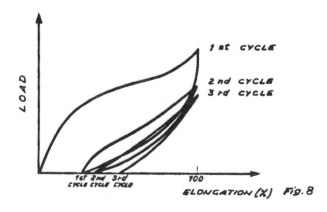

ELONGATION (%)   Fig.8

Our speculations are based on the following points and on
experimental evidence:
1) With particle dimensions in the 0.5-5 micron range and
at 23% volume PP, the layer or sheath of PP surrounding
any single soft rubber droplet is very thin on the order
of 0.05-0.5 micron.
2) The contact area between soft and hard domains is quite
large and the system is akin to an interpenetrated network.
Any deformation in tension of the continuous hard phase
will be accompanied by a deformation in compression of
the dispersed soft phase that will be "squeezed" in between the hard phase film-like layers.
3) PP/EPDM shows what we have called the "Pseudo Mullins
Effect" because of the close similarity to the Mullins effect observed in rubber vulcanizate reinforced with active
fillers (12), (13), (14). As can be seen in Figure 8 and
from the data presented in Table 7 the dynamically cured
PP/EPDM vulcanizates acquire a greater degree of elasticity due to the loss of PP domains reinforcing effect upon loading and unloading.
4) The "Pseudo Mullins Effect", just as the Mullins Effect
in rubber, is time and temperature dependent. The reinforcing effect of PP domains and the accompanying loss in
elasticity are restored in about 20 minutes at 120°C, 90
minutes at 80°C and 40 hours at 23°C.

Table 7: Effect of Three Extension-Retraction Cycles on
Elastic Properties of Thermoplastic Vulcanizate
4R (23% vol. PP) and Iso-crosslinked EPDM 4R
Reinforced with 23% vol. Carbon Black.

	Tensile (MPa)	Young Modulus (MPa)	Stored Energy 1st Cycle (%)	Stored Energy 3rd Cycle (%)
Thermoplastic 4R	65	14.5	35	60
Iso-crosslinked EPDM 4R - Carbon Black Reinforced	200	6.5	60	75

With the above points 1 to 4 in mind we have speculated
that when a PP/EPDM thermoplastic vulcanizate is strained
the load is taken up by the PP and most of the energy is
consumed uncoiling and aligning the PP chains surrounding the EPDM droplet while only a small portion of the energy is used in squeezing the EPDM domains. Once stress
is released, there is a partial recovery of the deformation as rubber domains regain their equilibrium dimension

dragging along the very thin highly flexible PP films.
Should the material be stressed again immediately thereaf-
ter, the load will be much lower since the PP chains had
already been aligned in the previous cycle. Nonetheless
the elastomer will be squeezed again much in the same way
storing up approximately the same energy so that deformation recovery will be almost complete and the hysteresis
loop will resemble that of a conventional elastomer. On
the other hand, if sufficient time elapses between two
successive cycles, the PP chain alignment disappears because the conditions of minimum free energy require this
to happen, thus, upon further elongation PP can once more
realign the chain resulting in the observation of the
stress-strain curve peculiar to PP/EPDMs.

ACKNOWLEDGEMENTS
The expert assistance of Mr. V. Braga in carrying out the
experimental work is gratefully acknowledged.

REFERENCES
1) G. Crespi, G. Di Drusco - Chim. Ind. Mi 48 731, 1967.

2) W. K. Fisher (to Uniroyal, Inc.), U.S. 3,758,643 (Sept.
   11, 1973); U.S. 3,835,201 (Sept. 10, 1974); U.S.
   3,862,102 (Jan. 21, 1975).

3) A. Y. Coran, B. Das and R. P. Patel (to Monsanto Co.)
   U.S. 4,130,535.

4) A. Y. Coran, R. Patel "Rubber Chemistry and Technology"
   Vol. 53 gg. 142-150.

5) A. Y. Coran and R. Patel, paper presented at the International Rubber Conference, Kiev, USSR, October 1978.

6) S. Danesi, E. Garagnani, Olefinic Thermoplastic Elastomers Kunststoffe, 37, Jahrgang, HEFT 3/84 Seiten 195-
   197.

7) A. M. Gessler (to Esso Res. & Engr. Co.) U.S. 3,037,954
   (June 5, 1962).

8) V. Braga, S. Giovanardi, E. Martini, Unpublished work.

9) V. Braga, S. Giovanardi, E. Martini, Unpublished work.

10) E. F. Cluff, E. K. Gladding and R. Pariser, J. Polymer
    Sci., 45, 341, 1960.

11) F. Milani, E. Martini, "Correlation between structure
    and low temperature properties of EPM and EPDM elasto-
    mers". Paper presented at the International Rubber Con-
    ference - Goteborg, June 1986.

12) F. Bueche, Physical Properties of Polymers, Intersci-
    ence, New York, 1962.

13) L. Mullins, Rubber Chem. Technol., 21, 281 (1948).

14) L. Mullins and N.R. Tobin, Rubber Chem. Technol., 30,
    555 (1957).

AUTHOR INDEX

SUBJECT INDEX

187